This book is to be returned on
or before the date stamped below

Acoustics of long spaces

Theory and design guidance

J. Kang

Published by Thomas Telford Publishing, Thomas Telford Ltd,
1 Heron Quay, London E14 4JD

URL: http://www.thomastelford.com

Distributors for Thomas Telford books are
USA: ASCE Press, 1801 Alexander Bell Drive, Reston, VA 20191-4400, USA
Japan: Maruzen Co. Ltd, Book Department, 3–10 Nihonbashi 2-chome, Chuo-ku, Tokyo 103
Australia: DA Books and Journals, 648 Whitehorse Road, Mitcham 3132, Victoria

First published 2002

A catalogue record for this book is available from the British Library

ISBN: 0 7277 3013 4

Typeset by Academic + Technical, Bristol
Printed and bound in Great Britain by MPG Books, Bodmin

Preface

The subject of this book is the acoustics of long spaces, such as road or railway tunnels, underground/railway stations, corridors, concourses and urban streets, where one dimension is much greater than the other two, although the other two are still relatively large compared to the acoustic wavelength. Acoustics is a major concern in many long spaces. For example, in urban streets and road tunnels noise pollution is a serious problem; in public buildings the noise disturbance between rooms through a corridor is often significant; and in underground/railway stations poor speech intelligibility of public address (PA) systems can cause loss of important travel information and misunderstanding of vital instructions during an emergency. An outstanding feature of long enclosures is that classic room acoustic theory is not applicable since the assumption of a diffuse field does not hold with the extreme dimensional condition.

This book attempts to provide a concise 'state-of-the-art' exposition of acoustics of long spaces. It presents the fundamentals of acoustic theory and calculation formulae for long spaces, and gives guidelines for practical design with extensively illustrated work. The book would be an addition to the existing acoustic books that mainly deal with regularly-shaped (i.e. quasi-cubic) spaces. It is hoped that the book is useful for practical design and also can serve as an up-to-date reference for research and teaching.

The book is relevant to several groups of readership. Related disciplines include acoustics, architecture, urban planning, building services, civil engineering, environmental engineering, transport engineering, mechanical engineering and electrical engineering. The book is reasonably self-contained. To help those readers who are not familiar with the field, an introductory chapter is included to provide some prior knowledge of elementary acoustics.

The book is divided into six chapters. Chapter 1 describes briefly fundamental concepts, basic theories and modelling techniques relating to general room acoustics. This includes physical properties of sound waves, behaviour of sound in front of a boundary, sound field in enclosures, computer simulation and scale modelling, and speech intelligibility. Chapter 2 presents theories and computer models for long spaces. The chapter begins with a brief discussion of the unsuitability of classic room acoustic theory for long enclosures; it then presents theories and formulae based on the image source method and the radiosity method, and reviews other relevant investigations. This is followed by a method for predicting train noise in underground stations and an overall model for practical prediction. In the next three chapters, design guidelines are presented with a large number of illustrations. Chapters 3 and 4, the former for long enclosures and the latter for urban streets, are based on parametric studies using the theoretical/computer models, whereas in Chapter 5 a series of physical scale-model measurements on the effectiveness of strategic architectural acoustic treatments are described. Particular attention is given to the improvement of the speech intelligibility of multiple loudspeaker PA systems in underground stations. Chapter 6 discusses speech intelligibility in long enclosures based on a series of articulation tests. Differences and implications of intelligibility between three languages are also analysed.

I wish to express my appreciation to Raf Orlowski of Arup Acoustics, Glenn Frommer of the Hong Kong Mass Transit Railway Corporation (MTRC) and Paul Richens of the University of Cambridge for their support on my recent research and consultancy work relating to this book. For assorted assistance, I must include mention of J. Murphy, D. J. Oldham, B. M. Shield, P. Barnett, J. L. Zhang, R. Carlyon, J. Owers, J. Rose, P. Tregenza, J. Till, K. Steemers, B. Gibbs, J. Z. Kang, M. Zhang, J. Picaut, K. M. Li, K. K. Iu, A. Dowling, J. Ffowcs-Williams, D. King, B. C. J. Moore, E. Milland, J. W. Edwards, L. N. Yang, P. Grasby, B. Everett, D. Crowther, M. Standeven, G. Moore, M. Trinder, J. Hodson, B. Logan, N. Craddock, P. Ungar, C. Woodhouse, J. Dubiel, L. Reade, C. Ovenston, R. Rajack, H. Day, N. Emerton, editors and referees of my publications relating to this book, and many colleagues at the University of Sheffield, the University of Cambridge, Fraunhofer-Institut für Bauphysik and Arup Acoustics. I also wish to thank my family for their unceasing support and encouragement. Much of the work was supported financially by the Hong Kong MTRC, the Cambridge Overseas Trust, the Lloyd's Foundation, the Humboldt Foundation and the Royal Society.

Contents

Notation

a	cross-sectional height
a_0	radius
A	total absorption in a room
$AC_{(l',n')(l,m)}$	form factor from emitter $A_{l',n'}$ to receiver $C_{l,m}$
b	cross-sectional width
c	sound speed in air
$C_k(t)_{l,m}$	a kth order patch source
d	distance from the receiver to a train section or to the tunnel entrance
d_i	distance from the image source i to the receiver
d_l, d_m, d_n	coordinate of the patch centre along the length, width and height
d_s	distance between the source and an end wall
$d_{(l',n'),(l,m)}$	mean beam length between two patches
dd_l, dd_m, dd_n	patch size along the length, width and height
D	sound attenuation ratio in dB/m
D_d	distance from the first train section to an end wall before deceleration
D_i	distance from the last train section to an end wall after acceleration
D_j	distance between receiver i and the jth source
D_0	average reflection distance
D_p	projection of D_0 into the image source plane
D_s	station length
$D_{z,m,n}$	distance between the receiver and image source (m,n)
$E(t)$	sound energy density at time t
$E_{z,m,n}$	sound energy from image source (m,n) between time t and $t+\Delta t$
f	frequency
f_{n_x,n_y,n_z}	eigenfrequency

f_0	resonant frequency
$f(\beta)$	radiation intensity of the source in direction β
$f(\eta_x, \eta_y)$	radiation intensity of the source in direction (η_x, η_y)
F	modulation frequency
G	number of image source planes
I	sound intensity
$I(\eta)$	sound energy intensity in direction η
j	source index
K_W	constant in relation to L_W
$l, l'm, m', n, n'$	patches along the length, width and height
l_1	width of boundary i
l_2	width of the boundary which is vertical to boundary i
l_3	distance between the receiver and the cross-section with P_{IN}
l_T	length of the train section
L	length of a long enclosure or a street canyon
L_{D_j}	steady-state sound pressure level (SPL) at receiver i caused by the jth source
L_{eq}	equivalent continuous sound level
L_i	steady-state SPL at the receiver i caused by all the sources
L_0	length of the model train
L_p	sound pressure level
Lp_D	SPL difference between model train and actual train
L_T	length of an actual train
L_W	sound power level
L_{ref}	reference SPL
L_z	steady-state SPL at a receiver with a distance of z from the source
$L(t)_z$	energy response at a receiver with a distance of z from the source
$L(t)_{z,\psi}$	energy response at receiver (z, ψ)
m	modulation index
$m(F)$	modulation transfer function
m, n	order of image sources
$M, \mathbf{M}$	air absorption in Np/m and dB/m, respectively
n	position of a train section in a train
n_x, n_y, n_z	positive integers, one or two of them may be zero
N	approximate number of image sources
N_0	number of train sections in the model train
$N_R + 1$	number of receivers between two loudspeakers
N_s	number of multiple sources
N_T	number of train sections in an actual train
N_X, N_Y, N_Z	number of patches along the length, width and height
p	significance level

p	sound pressure
$p(t)$	impulse response
P	sound power of the source
P_{IN}	input power
P_{SO}	power flow at the receiver
$P_{SO,i}$	power flow at the receiver caused by boundary i
q_x, q_y, q_z	ratio between adjacent patches
Q	source directivity factor
r	correlation coefficient
r	distance between the receiver and the train entrance
$r(t)$	squared impulse response
R	average order of image sources
$R_{l,m}$	mean beam length between a patch and the receiver
R_m	receiver with a distance of $\sigma/2$ to R_0
R_0	receiver in the same cross-section with the central source
R_T	total room constant
$R(R_x, R_y, R_z)$	receiver
RT_z	reverberation time at the receiver with a distance of z from the source
S	cross-sectional area
S_i	area of surface i
$S_{l,m}$	mean beam length from a patch to the source
S_n	SPL caused by the nth train section at receiver R
S_0	total surface area
S_T	train noise level
$S(n, v)$	difference in L_W between the nth train section and the reference section
$S(S_x, S_y, S_z)$	sound source
$Sss(d)$	SPL with both the train section and the receiver in the station
$Sts(d)$	SPL with the train section in the tunnel and the receiver in the station
$Stt(d)$	SPL with the train section in the tunnel and the receiver at the end wall
ΔS_j	difference in L_W between the jth source and the reference source
t, t'	time variable
t_j	time delay of the jth source
t_0	estimated reverberation time
$t_{z,m,n}$	arrival time of the image source (m, n)
T_a	time range before deceleration
T_d	duration of deceleration
T_e	time range after deceleration

T_i	duration of acceleration
T_j	reverberation time (RT) at receiver i resulting from the jth source
T_s	duration when the train is static at the station
u	sound energy density
U	cross-sectional perimeter
v	train velocity
v'	reference train velocity
V	room volume
V_d	train velocity before deceleration
V_i	train velocity after acceleration
W	sound power
X_n	position of the nth train section at time t
X_0	position of the first train section at time t
z	source-receiver distance
α	absorption coefficient
$\bar{\alpha}$	mean absorption coefficient
$\alpha_C, \alpha_A, \alpha_B$	absorption coefficient of the ceiling and side walls (façades)
α_F, α_G	absorption coefficient of the floor and ground
$\alpha_e, \alpha_U, \alpha_V$	absorption coefficient of end walls
α_i	absorption coefficient of boundary i
α_n	normal absorption coefficient
β	angle between the long axis and the source radiation
θ	angle between the boundary normal and an incident ray
λ	wavelength
ρ	reflection coefficient
ρ_e	reflection coefficient of an end wall
ρ_0	density of air
ρ_θ	angle-dependent reflection coefficient
σ	source spacing
ψ	source-receiver distance along the width
$si(y), ci(y)$	$-\int_y^\infty (\sin t/t)\, \mathrm{d}t, \ -\int_y^\infty (\cos t/t)\, \mathrm{d}t$

Acronyms

AI	articulation index
ACL	acoustics of long spaces (a computer model)
BEM	boundary element method
CLC	combined level change
DI	directivity index
EDT	early decay time
LD	level decrease
LI	level increase
MTF	modulation transfer function
MTRC	mass transit railway corporation (Hong Kong)
MUL	multiple sources in long spaces (a theoretical/computer model)
PA	public address
PB	phonetically balanced
RASTI	rapid speech transmission index
RT	reverberation time
RT30	reverberation time obtained from $-5\,dB$ to $-35\,dB$
S/N	signal to noise (ratio)
SPL	sound pressure level
STI	speech transmission index
TNS	train noise in stations (a computer model)

1. Basic acoustic theories and modelling techniques

This chapter briefly describes fundamental concepts, basic theories and modelling techniques relating to general room acoustics. It begins with a review of physical properties of sound waves, followed by the behaviour of sound in front of a boundary, including absorption, diffusion and reflection. It then discusses the sound field in enclosures in terms of wave acoustics, geometrical acoustics and statistical acoustics. This is followed by a brief review of computer simulation and scale modelling in room acoustics. The final Section of this chapter outlines methods for assessing speech intelligibility.

1.1. Basic theory of sound

1.1.1. Sound and the basic properties of a sound wave

Sound may be defined in general as the transmission of energy through solid, liquid or gaseous media in the form of vibrations. Two essential factors for sound energy to exist are sound source, i.e. a vibrating body, and a medium.

In a medium, each vibrating particle moves only an infinitesimal amount to either side of its normal position. It is first displaced in the direction of propagation of the wave, then it will move back to its undisturbed position and continue towards a maximum negative displacement, due to the action of the rarefaction. The time for completing a full circuit by a displaced particle is called the period, T. Usually the oscillations are repeated and the repetition rate is described by the reciprocal of the period, i.e. frequency, f. In other words, frequency is the number of oscillations per second. The unit of frequency is Hertz (Hz). The period is related to the frequency by

$$T = \frac{1}{f} \tag{1.1}$$

The continuous oscillation of the source propagates a series of compressions and rarefactions outwards through the medium. The distance between adjacent regions where identical conditions of particle displacement occur is called the wavelength. It is the distance a sound wave travels during one cycle of vibration.

The velocity with which sound travels through air varies directly with the equilibrium air pressure and inversely with the equilibrium air density. At standard pressure (760 mm Mercury) and 20°C, it is approximately 340 m/s.

There is a relationship between the wavelength, λ, the velocity of propagation, c, and the frequency, namely

$$c = f\lambda \tag{1.2}$$

A simple sound source can be considered as a rigid piston in a very long tube, as demonstrated in Figure 1.1. As the piston moves forward and backwards from the equilibrium position, a plane wave is created along the tube. A plane wave is a special case in which the acoustic variables are functions of only one spatial coordinate.

Another simple sound source may be considered as a pulsating sphere, alternatively increasing and decreasing its diameter. Such a source radiates sound uniformly in all directions and is called a spherical source. In a free field, where sound can travel to any direction without obstructions, sound waves from a spherical source are virtually spherical (see Figure 1.1).

1.1.2. Sound pressure, sound intensity and sound power

Sound can be sensed by the measurement of some physical quantity in the medium that is disturbed from its equilibrium value. The sound pressure is a commonly used index. The instantaneous sound pressure at a point is the incremental change from the static pressure at a given instant caused by the presence of a sound wave. The effective sound pressure at a point is the root-mean-square (rms) value of the instantaneous sound pressure over a time interval at that point. For periodic sound pressures, the interval should be an integral number of periods. The unit of sound pressure is N/m^2. Sound pressures are extremely small. At a distance of a metre from a talker, the average pressure for normal speech is about $0.1\,N/m^2$ above and below atmospheric pressure, whereas the atmospheric pressure is about $1.013\,N/m^2$ at sea level.

The sound intensity is measured in a specified direction. It is the average rate at which sound energy is transmitted through a unit area perpendicular to the specified direction. The unit of sound intensity is W/m^2. In a free-progressive plane or spherical sound wave the intensity in the direction of propagation is

$$I = \frac{p^2}{\rho_0 c} \tag{1.3}$$

where p is the rms pressure of the sound and ρ_0 is the ambient density of the medium.

The sound pressure in a medium is naturally related to the power of the source of sound. The sound power of a source is the rate at which acoustic

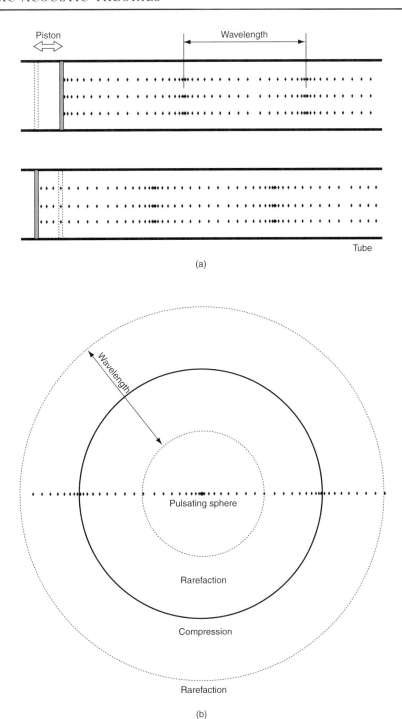

Figure 1.1. *(a) Plane wave and (b) spherical wave*

energy is transferred from a vibrating source to a medium. This power is measured in watts. The sound energy density is the sound energy in a given infinitesimal part of the gas divided by the volume of that part of the gas. The unit is W/m^3.

In air, sound is transmitted in the form of a longitudinal wave, i.e. the particles are displaced along the direction of propagation of the wave. In a plane wave the pressure and particle velocity have the same phase.

1.1.3. Levels and decibels

It is known that the ear does not respond linearly to sound intensity or pressure. Perceived changes in intensity tend to be proportional to the ratios between the intensities concerned. Thus, it is more convenient to use a logarithmic unit to measure intensity — the decibel (dB). Decibels can be found by the following formula

$$L_I = 10 \log \left(\frac{I}{I_0} \right) \tag{1.4}$$

where L_I is the sound intensity level (dB), I is the intensity of a sound (W/m^2) and I_0 is the reference intensity, $10^{-12} \, W/m^2$. I_0 is the minimum sound intensity audible to the average human ear at 1000 Hz.

Another advantage of using logarithmic scale is that it allows the huge range of human sensitivity to be conveniently represented by a scale of small numbers. From the threshold of audibility to the threshold of pain, the intensity ratio is about 10^{12}.

Similarly, the sound pressure level (SPL), L_P, is defined as

$$L_P = 10 \log \left(\frac{p^2}{p_0^2} \right) \tag{1.5}$$

or

$$L_P = 20 \log \left(\frac{p}{p_0} \right) \tag{1.6}$$

where p_0 is the reference pressure, $2 \times 10^{-5} \, N/m^2$. It is chosen so that the numerical values for intensity and pressure are approximately the same at standard atmospheric conditions.

The sound power level of a source is given by

$$L_W = 10 \log \left(\frac{W}{W_0} \right) \tag{1.7}$$

where W is the sound power (W) and W_0 is the reference sound power, $10^{-12} \, W$.

Very often contributions from more than one sound source are concerned. If the phases between sources of sound are random, the sounds can be added together on a linear energy (pressure squared) basis. To add SPLs, the mean-square sound pressure from each source should be first determined according to equation (1.5). The total SPL can then be obtained from the summation of the mean-square sound pressures. Assume there are n sources and L_{pi} is the SPL of each source. According to the above procedure, the total SPL, L_p, can be calculated by

$$L_P = 10 \log \left(\sum_{i=1}^{n} 10^{L_{pi}/10} \right) \qquad (1.8)$$

1.1.4. Inverse-square law and source directionality

If a spherical source is very small, say the source radius is small compared with one-sixth wavelength, the source is called a point source. In a free field, at a distance z from the centre of a point source, the intensity is the sound power of the source divided by the total spherical area of the sound wave at z

$$I = \frac{W}{4\pi z^2} \qquad (1.9)$$

In other words, the intensity at any point is inversely proportional to the square of its distance from the source (inverse-square law). This is equivalent to a reduction of 6 dB SPL at each doubling of the distance from the source.

When the dimensions of a sound source are much smaller than a wavelength, the effect of the source shape on its radiation will be negligible, providing all parts of the radiator vibrate substantially in phase. However, most sources are not small compared with the wavelength, or there are reflecting obstacles nearby that disturb their sound field. Thus, it is necessary to consider their directional properties. The directivity factor, Q, is defined as the ratio of the intensity at some distance and angle from the source to the intensity at the same distance if the total power from the source were radiated uniformly in all directions. The directivity index, DI, is defined as

$$DI = 10 \log Q \qquad (1.10)$$

For practical applications it is more convenient to express DI in terms of the SPL

$$DI = L_{p\theta} - \bar{L}_p \qquad (1.11)$$

where $L_{p\theta}$ is the SPL at the direction of interest, and $\bar{L}_p$ is the average SPL in all directions. Both $L_{p\theta}$ and $\bar{L}_p$ are determined for the same fixed distance from the source.

1.1.5.   Frequency analysis and frequency weighting

Frequencies in the audio range are from about 20 to 20 000 Hz. Because most sounds are complex, the SPL can be measured in a series of frequency intervals called frequency bands. Octave and fractional octave bands are most commonly used. In each octave band, the upper limiting frequency is exactly twice the lower limiting frequency. The centre frequency is the geometric mean of the upper band limit and lower band limit. The centre frequencies that have been standardised for acoustic measurements are 31·5, 63, 125, 250, 500, 1000, 2000, 4000, 8000 and 16 000 Hz. One-third octave bands are formed by dividing each octave band into three parts. Successive centre frequencies of one-third octave bands are related by factor $\sqrt[3]{2}$. For example, the one-third octave bands contained within the octave centred on 500 Hz will have centre frequencies 400, 500 and 630 Hz.

The relationships between SPL and frequency are often required for acoustic analysis. Data so plotted are called a sound spectrum. Figure 1.2 shows two typical spectra.

When making measurements on a sound source it is often considered desirable to have a single reading. Frequency weighting takes typical human response to sound into account when the sound level in each frequency is adjusted. The adjusted levels are then added to produce a single number in decibels. Standard weighting networks include A, B, C and D, and the resultant decibel values are called dB(A), dB(B), etc. The A-weighting network, which is commonly used in noise legislation, was originally designed to approximate the response of the human ear at relatively low sound levels. It tends to considerably neglect low-frequency sound energy.

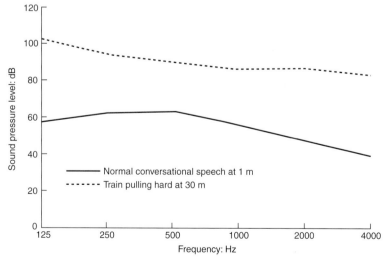

Figure 1.2. Two typical spectra [1.1]

The sound level at a given receiver is the reading in decibels of a sound-level meter. The meter reading corresponds to a value of the sound pressure integrated over the audible frequency range with a specified frequency weighting and integration time.

1.2. Sound field in front of a boundary

1.2.1. Absorption, reflection and transmission coefficient

When sound waves fall on a boundary, their energy is partially reflected, partially absorbed by the boundary and partially transmitted through the boundary, as illustrated in Figure 1.3.

The absorption coefficient, α, is the ratio of the sound energy that is not reflected from the boundary to the sound energy incident on it. It is a function of frequency, and may take on all numerical values between 0 and 1. When the sound wave is incident under an angle θ to the normal, the absorption coefficient is called the oblique-incidence absorption coefficient, given as α_θ. The normal-incidence absorption coefficient corresponds to $\theta = 0$. When the incident sound is evenly distributed in all directions, the absorption coefficient is called the random-incidence absorption coefficient or statistical absorption coefficient. The absorption characteristics of a material can be measured using an impedance tube or a reverberation room. The impedance tube method gives the normal-incidence absorption coefficient, and the coefficient determined in a reverberation room is termed the Sabine absorption coefficient, which is usually close to the random-incidence absorption coefficient.

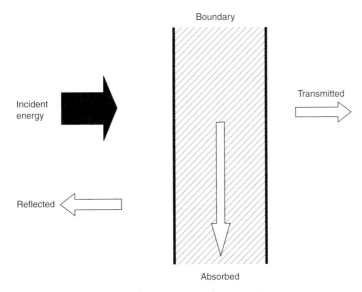

Figure 1.3. Sound reflection, absorption and transmission

The reflection coefficient ρ is the ratio of the sound energy that is reflected from the boundary to the sound energy incident on it, i.e. $\rho = 1 - \alpha$. The fraction of incident energy that is transmitted through the boundary is called the transmission coefficient. Sound absorption, reflection and transmission coefficients are all frequency dependent.

1.2.2. Sound reflection, diffusion and diffraction

When sound waves strike a boundary, if the boundary is acoustically rigid and smooth, the angle of incidence of the wavefront is equal to the angle of reflection. In other words, a sound wave from a given source, reflected by a plane surface, appears to come from the image of the source in that surface, as illustrated in Figure 1.4. The reflection pattern from curved surfaces can be similarly determined by the simple application of geometrical laws.

If there are considerable irregularities on a reflecting boundary, and the sizes of which are comparable with the wavelength or smaller than it, the incident sound energy will be scattered to a particular solid angle. If the directional distribution of the reflected or the scattered energy does not depend in any way on the direction of the incident sound, the boundary is called a diffusely reflecting boundary.

If the incident wavelength is not small compared to the reflecting surface, diffraction must be considered. Diffraction is the bending or spreading out of a sound wave after it intersects an aperture, a solid object, a recess or a surface protrusion. Diffraction is more marked for low frequencies than for high frequencies. It has been shown that when the minimum dimension of a

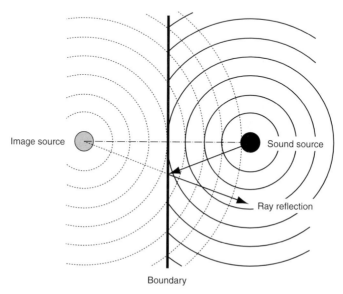

Figure 1.4. Reflection of a sound wave from a plane surface

panel is about ten times the wavelength some diffraction occurs and at less than about five times the wavelength the incident energy is diffracted [1.2].

1.2.3. Sound absorption

Absorptive materials and constructions can be divided into three fundamental types, namely porous absorbers, panel and membrane absorbers, and resonator absorbers.

The porous absorber is characterised by its pores that are accessible from the outside. When sound waves impinge on a porous material, part of the sound energy is converted into thermal energy due to viscous flow losses caused by wave propagation in the material and internal frictional losses caused by motion of the material's fibres. The absorption characteristics of a porous material are dependent upon a number of variables, including its thickness, density, porosity, flow resistance and fibre orientation. Generally speaking, the absorption is large at high frequencies and the performance at low frequencies depends mainly on the thickness and the flow resistance. Since the maximum sound absorption will occur when the particle velocity is at a maximum, namely, at a distance of $\lambda/4$, $3\lambda/4$, etc., from a rigid backing wall, a material needs to be rather thick if it is required to absorb sound energy at low frequencies. Alternatively, it may be mounted at some distance from the rigid wall. The absorption coefficient of a typical porous material is shown in Figure 1.5.

A panel or membrane can be set into vibration when it is stricken by a sound wave. Usually the panel or membrane is mounted at some distance from a rigid wall, and the combination behaves analogously to a mass-spring system. Owing to the friction in the panel or membrane itself, in its supports and in any space behind it, an energy loss occurs and hence some sound absorption takes place. The resonant frequency of such a system depends mainly on the stiffness, surface density, thickness and elastic modulus of the material, and the depth of the airspace behind. In practice, panel and membrane absorbers are most useful at low to mid frequencies. The absorption coefficient of a typical panel absorber is shown in Figure 1.5. The frequency range of absorption can be extended by placing a porous material in the airspace because this will provide extra damping (friction).

A single resonator comprises a cavity within a massive solid, connected to the air outside through a restricted neck. The impinging sound energy causes the air in the neck to vibrate in the resonant frequency range and, thus, some sound energy is dissipated by means of viscous losses. A single resonator is only effective for a narrow range of frequency, and is most efficient at low frequencies.

A perforated panel mounted at some distance from a rigid wall can be regarded as a series of single resonators. The perforation may be in the form

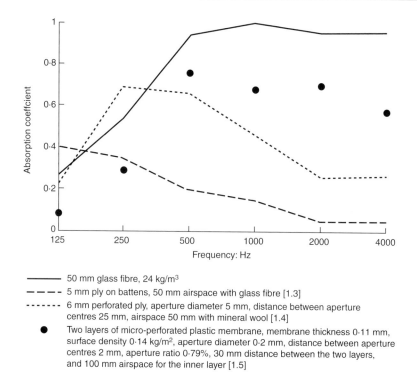

Figure 1.5. Absorption coefficient of typical absorbers

of holes or slits. The resonant frequency and the frequency range of absorption of such absorbers, depend mainly on the aperture diameter, aperture spacing, panel thickness and the depth of the airspace. When there is a porous material behind the perforated panel to provide extra damping, the frequency range of absorption can be considerably extended, although the maximum absorption at resonant frequencies is normally lowered. Figure 1.5 shows the absorption coefficient of a typical perforated panel absorber.

The micro-perforated absorber is a cavity backed panel with low perforation ratio but many apertures of sub-millimetre size. In comparison with conventional perforated panel absorbers, an outstanding feature of micro-perforated absorbers is that the acoustic damping of the apertures becomes significant when the apertures are very small. Consequently, for micro-perforated absorbers it is not necessary to provide extra damping using porous materials [1.5,1.6]. Owing to this feature, the absorber can be used in some special situations. For example, it can be made from transparent materials like plastic glass.

Combinations of the above fundamental types of absorber are also widely used. For example, multiple layers of panel or perforated panel may have multiple resonant frequencies and, thus, the frequency range of absorption can be broadened in comparison with a single layer. The absorption performance of a micro-perforated membrane backed by an airspace can be rather good in a wide range of frequency due to the resonance from both membrane and apertures [1.5]. The absorption coefficient of such a structure is shown in Figure 1.5.

Functional absorbers are three-dimensional units of sound absorbing material that are suspended freely in a room space with some distance from the boundaries. Since sound energy is free to impinge on all sides of these units, they provide a powerful absorbing effect.

1.3.   Sound in enclosures

For small rooms of simple shape, the interior sound field can be described in precise mathematical terms by considering the physical wave nature of sound. For large irregularly shaped rooms, such a description becomes very difficult but a statistically reliable statement can be made of the average conditions in the room.

1.3.1.   Room resonance

If sound is supplied to a closed tube and the diameter of the tube is small compared with the wavelength, the tube will 'resonate' at certain frequencies where the tube length is an integral multiple of a half wavelength. The forward- and backward-travelling waves add in magnitude to produce what is called a standing wave. The frequencies are called resonant frequencies, natural frequencies, normal frequencies, or eigenfrequencies.

Similar standing waves can also be set up in a room. The standing waves travel not only between two opposite, parallel boundaries, but also around the room involving the boundaries at various angles of incidence. For a rectangular enclosure, the frequencies of these standing waves are given by

$$f_{n_x, n_y, n_z} = \frac{c}{2} \left[\left(\frac{n_x}{a} \right)^2 + \left(\frac{n_y}{b} \right)^2 + \left(\frac{n_z}{L} \right)^2 \right]^{1/2} \tag{1.12}$$

where f_{n_x, n_y, n_z} is the resonant frequency, a, b and L are the dimensions of the enclosure, and n_x, n_y and n_z are positive integers; one or two of them may also be zero. The values found from equation (1.12) correspond to the room modes. They can be divided into three categories: axial modes, two n are zero and the waves travel along one axis; tangential modes, one n is zero and the waves are parallel to one pair of parallel boundaries and are obliquely incident on two other pairs of boundaries; and oblique modes, no n is zero and the waves are obliquely incident on all boundaries.

The sound pressure is proportional to the product of three cosines, each of which describes the dependence of the pressure on one coordinate

$$p_{n_x, n_y, n_z}(x, y, z) \propto \cos \left(\frac{n_x \pi x}{a} \right) \cos \left(\frac{n_y \pi y}{b} \right) \cos \left(\frac{n_z \pi z}{L} \right) \tag{1.13}$$

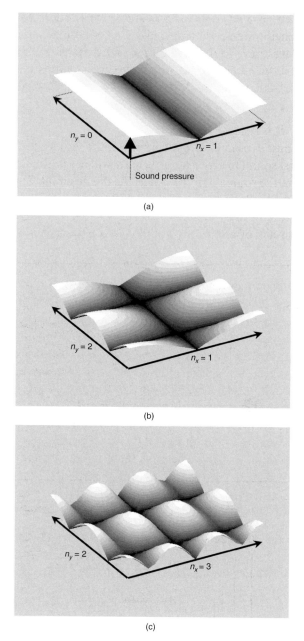

*Figure 1.6. Distribution of sound pressure on a section through a rectangular room:
(a) axial mode (1, 0, 0); (b) tangential mode (1, 2, 0); (c) tangential mode (3, 2, 0)*

Based on equation (1.13), the pressure distribution with three typical room
modes is shown in Figure 1.6.

If a room is large and irregular in shape, in the audio-frequency range the
number of room modes will be enormous and their pressure distribution will be

very complex. Thus, it is very difficult to treat so many standing waves individually. Since the wavelengths are short compared with room dimensions, it is reasonable to treat a sound wave as a sound ray. In other words, the sounds behave like light rays constantly reflected and re-reflected between boundaries, and the physical wave nature of sound, such as diffraction and interference, is ignored.

1.3.2. Sound at a receiver

Consider the acoustic response at a listening point to a short burst of sound from a source. The first sound to arrive at the listener will be the sound that travels in a straight line from the source. It is known as the direct sound. This is followed by a series of sounds that have travelled by paths including one or more reflections from room surfaces, as illustrated in Figure 1.7(a). In comparison with the direct sound, the amplitude of a reflected sound is less because it travels farther and,

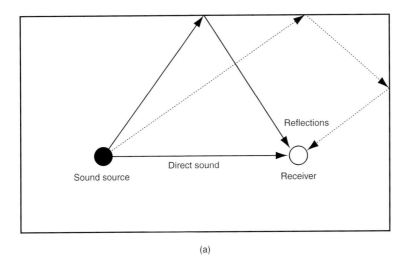

(a)

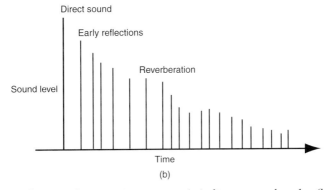

(b)

Figure 1.7. Reflection diagram in a room: (a) direct sound and reflections; (b) echogram

also, part of the sound energy is absorbed by the reflected surfaces. The whole process can be represented as a diagram of sound level against time, as shown in Figure 1.7(b). Such a diagram is known as an echogram.

Reflections that arrive at the listener immediately after the direct sound are called early reflections. After a certain period, the number of reflections becomes so high that individual reflections are no longer distinguishable. These late reflections are called the reverberant sound.

1.3.3. Reverberation

After a source starts to supply sound energy in a room, a certain period is needed to build up an equilibrium sound level. Similarly, after the sound source stops, the sound will still be audible for sometime. This decay process is called the reverberation of the room. The reverberation time is defined as the time taken for a sound to decay 60 dB after the source is stopped. To determine the reverberation time, a decay curve is used. It is a plot of the decay of SPL in a room as a function of time after the source of sound has ceased.

Sabine found that reverberation was a function of the volume of the room and of the amount of sound absorption within it [1.7]

$$RT = \frac{0 \cdot 16V}{A} \tag{1.14}$$

where RT is the reverberation time (s), V is the volume of the room (m^3), and A is the total absorption in the room (m^2). A is found by multiplying each individual surface area S_i by its absorption coefficient α_i and adding the whole together

$$A = \sum_{i=1}^{n} \alpha_i S_i \tag{1.15}$$

The Sabine formula is applicable for live rooms with low absorption but becomes progressively more inaccurate as the average absorption coefficient increases. The Eyring formula is a modification of the Sabine formula

$$RT = \frac{0 \cdot 16V}{-S_0 \ln(1 - \bar{\alpha})} \tag{1.16}$$

where S_0 is the total surface area (m^2) and $\bar{\alpha} = A/S_0$ is the mean absorption coefficient.

At high frequencies, the absorption of sound by air is significant and should be taken into account

$$RT = \frac{0 \cdot 16V}{-S_0 \ln(1 - \bar{\alpha}) + 4MV} \tag{1.17}$$

where M is the energy attenuation constant due to air absorption (Np/m). It is dependent on relative humidity and air temperature [1.8].

The above reverberation formulae embody the assumption of a diffuse sound field, where acoustic energy is uniformly distributed throughout an entire room, and at any point the sound propagation is uniform in all directions. Although there is no real sound field that strictly meets this condition, the Sabine and Eyring formulae are accurate enough for many rooms.

The reverberation time can be obtained from interrupted noise or from impulse response $p(t)$ [1.9]. Correspondingly, the decay curves may be either measured after the actual cut-off of a continuous sound source in a room or derived from the reverse-time integrated squared impulse response of the room. In an ideal situation with no background noise, the integration should start at the end of the impulse response and proceed to the beginning of the impulse response. Thus, the energy decay $E(t)$ as a function of time t is

$$E(t) = \int_t^\infty p^2(\iota)\,d\iota = \int_\infty^t p^2(\iota)\,d(-\iota) \qquad (1.18)$$

Reverberation time is usually determined using the rate of decay given by the linear regression of the measured decay curve from a level 5 dB below the initial level to 35 dB below. It is called the RT30. Since the subjective judgement of reverberation is well correlated to the early slope of the decay curve, the early decay time (EDT) is often used. It is obtained from the initial 10 dB of the decay. For both the RT30 and EDT, the slope is extrapolated to correspond to a 60 dB decay. In a diffuse sound field, a decay curve is perfectly linear and, thus, the RT30 and EDT should have the same value.

1.3.4. Steady-state sound energy distribution

After a sound source in a room is turned on, energy is supplied to the room by the source at a rate faster than it is absorbed by the boundaries and air until a condition of equilibrium is achieved. In a diffuse field the steady-state sound energy distribution can be calculated by [1.10]

$$L_z = L_W + 10\log\left(\frac{Q}{4\pi z^2} + \frac{4}{R_T}\right) \qquad (1.19)$$

where $R_T = S_0\alpha_T/(1-\alpha_T)$ and $\alpha_T = \bar{\alpha} + 4MV/S_0$. L_W is the sound power level of the source, Q is the directivity factor of the source, z is the source-receiver distance, and R_T is called the total room constant.

In equation (1.19) the sound field is divided into two distinct parts, the direct sound field and the reverberant sound field. They are represented by $Q/4\pi z^2$ and $4/R_T$, respectively. The direct sound decreases 6 dB for every doubling of distance, whereas the reverberant sound is constant throughout a room. The distance with $Q/4\pi z^2 = 4/R_T$, namely the direct and reflected components are the same, is known as the reverberation radius. In a diffuse field the SPL becomes approximately constant beyond the reverberation radius.

1.4. Modelling techniques

Since most room acoustic problems cannot be resolved by purely analytical procedures, computer simulation and scale modelling are commonly used. This Section briefly reviews the modelling techniques.

1.4.1. Computer simulation

Computer simulation in room acoustics started from the 1960s [1.11], and the techniques have been developed continuously ever since. A number of different modelling techniques has been developed, including the image source method [1.12,1.13], ray tracing [1.14,1.15], various forms of beam tracing [1.16,1.17], the radiosity method [1.18–1.20], the finite element method (FEM) and the boundary element method (BEM).

The image source method treats a flat surface as a mirror and creates an image source. The reflected sound is modelled with a sound path directly from the image source to a receiver. Multiple reflections are achieved by considering further images of the image source. At each reflection the strength of the image source is reduced due to the surface absorption. With the image source method the situation of a source in an enclosure is replaced by a set of mirror sources in a free field visible from the receiver considered. The acoustic indices at the receiver are determined by summing the contribution from all the image sources. A disadvantage of the image source method is that it slows exponentially with increasing orders of reflection as the number of images increases. In addition, validity and visibility tests are required for image sources.

A sound ray can be regarded as a small portion of a spherical wave with vanishing aperture, which originates from a certain point. Ray tracing creates a dense spread of rays, which are subsequently reflected around a room and tested for intersection with a detector (receiver) such as a sphere or a cube. An echogram can be constructed using the energy attenuation of the intersecting rays and distances travelled. Particle tracing uses similar algorithms to ray tracing but the method of detection is different. With the particle model, the longer a particle stays in the detector, the higher its contribution to the energy density. Beams are rays with a non-vanishing cross-section. The beams may be cones with a circular cross-section or pyramids with a polygonal cross-section. By using beams, a point detector can be used. Beams are reflected around a room and tested for illumination of the detector.

The radiosity method is useful for considering diffusely reflecting boundaries. The method, also called radiation balance, radiation exchange or radiant interchange, was first developed in the nineteenth century for the study of radiant heat transfer in simple configurations [1.21]. Computer implementations of the method began almost as soon as computers were invented but computer graphics research has given the technique an enormous boost over the past ten years with the advent of more powerful computing resources [1.22,1.23]. In computer graphics, radiosity is used predominantly to calculate light energy. By considering relatively high frequencies, the method can also be used in the field of acoustics. A significant feature of this application is that the reverberation, or in other words, the time factor, must be taken into account. This can substantially increase the computation time. Basically, the radiosity method divides boundaries in a room into a number of patches (i.e. elements) and replaces the patches and receivers with nodes in a network. The sound propagation in the room can then be simulated by energy exchange between the nodes. The energy moving between pairs of patches depends on a form factor, which is defined as the fraction of the sound energy diffusely emitted from one patch that arrives at the other by direct energy transport.

Continuous efforts are being made to improve the simulation accuracy and efficiency. In addition to the basic methods described above, combined algorithms have been developed [1.16,1.19,1.24,1.25]. Considerable attention has been given to the simulation of diffuse reflection as well as the definition and measurement of diffusion coefficient and scattering coefficient. Another attempt is to include interference in the modelling process [1.26]. At each reflection of a sound wave from a boundary, a complex reflection coefficient is used to take the amplitude and phase change into account.

Acoustic finite element method and boundary element method are based on the approximation to the wave equation. The methods can model resonances in the frequency domain and wave reflections in the time domain. They have been successfully applied in acoustic simulation of small rooms, namely, the wavelength is larger or at least of the same order as the room dimensions [1.27]. With the development of more powerful computers, the application range can be extended to relatively large rooms [1.28]. The methods have also been used in simulating the sound field in vehicle passenger spaces and in urban streets [1.29].

1.4.2.   Physical scale modelling

Acoustic physical scale modelling has been applied to design and research work in architectural and building acoustics for several decades [1.30,1.31]. In comparison with computer simulation, a notable advantage of using scale modelling is that some complex acoustic phenomena can be considered, such as the diffraction behaviour of sound when it meets obstacles. Since the speed of sound in air is constant, and the sound behaviour when it hits an obstacle is determined by the relationship between the size of the object and the wavelength of sound, both time and wavelength should be scaled. In other words, in a $1:n$ scale model, the measured time factors should be enlarged by n times, and the frequency should be n times higher. The sound levels are not subject to any scaling. As to the model size, a wide range of scale factors from $1:2$ to $1:100$ has been used.

For model materials, although it is ideal to accurately simulate the boundary impedance, it is practically sufficient to model the absorption coefficient. Considerable data have been published on the absorption characteristics of materials at model frequencies. For example, acoustically hard surfaces can be reproduced using varnished timber or plastic glass, and windowpanes can be simulated with thin aluminium panels. To determine the absorption coefficient at model frequencies, a model reverberation room can be used.

A problem that arises in scale models is that the air absorption becomes very large in the ultrasonic range. Air absorption is due to both molecules of oxygen O_2 and moisture H_2O. By filling a scale model with dry air of relative humidity 2–3% or with oxygen-free nitrogen, it is possible to obtain similar air absorption at model and full scale. If the measurement data are processed using a computer, the excessive air absorption can be numerically compensated [1.32].

Many of the objective measurements that can be carried out at full scale can also be made at model scale. These include impulse response, sound distribution and decay process. The sound source can be a spark pulse or a small loudspeaker, and the receiver can be a small condenser microphone. Subjective testing can also be undertaken in relatively large models of $1:8$ or $1:10$ scale. Dry music or speech recorded in an anechoic chamber is played back in a scale model at an increased tape velocity and the recorded signal in the model is slowed down to be listened to over headphones. Moreover, by measuring impulse responses in both ears of a dummy head using small microphones, binaural auditory tests can be performed by convoluting dry music or speech with both impulses [1.33].

Water and light models have also been used to simulate acoustic phenomena since there are some common properties between sound waves, water waves and light waves. Water model experiments were carried out by Scott Russell as early as 1843 [1.34]. The models are useful for demonstration because the wave velocity is relatively slow. In place of sound, ripples on a shallow water basin are used. The practical use of water models, however, is restricted because they are only two-dimensional and the wavelength range to be handled is rather narrow.

The use of light models in acoustics is limited to very high frequency since the wavelengths of light are very small in comparison with room dimensions. Sound absorption can be simulated by light absorption, and diffusely reflecting boundaries can also be modelled. Lasers can be used to simulate sound rays. Light models are useful for examining first- or second-order reflections, as well as for investigating sound intensity distribution using luminous intensity. Their disadvantage, however, is that it is difficult to obtain information about reflection arrival times and reverberation.

1.5. Speech intelligibility

Speech intelligibility is an essential index in many spaces. This Section reviews three methods for assessing the intelligibility.

1.5.1. Articulation test

A quantitative measure of the intelligibility of speech may be obtained using an articulation test. Typically, a speaker reads lists of phonetically balanced (PB) syllables, words or sentences to a group of listeners, and the percentage of items correctly recorded by these listeners is called the articulation score [1.35,1.36]. The percentage of correctly received phrases is called speech intelligibility. In a number of languages the relationships between syllable scores, word scores and sentence intelligibility have been established.

1.5.2. Articulation index

The articulation index (AI) is a longstanding method for rating the effect of background noise on speech intelligibility [1.37]. To determine the AI, the signal-to-noise (S/N) ratio in each of the 20 one-third octave bands from 200 Hz to 5000 Hz is measured first. The 20 S/N ratios are then individually weighted by an importance function dependent on frequency. The weighted values are combined to give a single overall value, the AI. Further corrections include for reverberation time and for very high background levels. The AI uses a scale from 0 to 1, where 0 is total unintelligibility and 1 is equivalent to 100% intelligibility.

1.5.3. Speech transmission index

The speech transmission index (STI) is a commonly used index for intelligibility. It is based on the modulation transfer function (MTF) between source and receiver. The concept of the MTF was introduced into room acoustics by Houtgast and Steeneken as a measure for assessing the effect of a room on intelligibility [1.38]. The MTF can be described by treating a room as a signal transmission path [1.38–1.41]. For an input signal with a varying intensity $I_i(t) = \overline{I_i}(1 + \cos 2\pi Ft)$, the output signal takes the form $I_o(t) = \overline{I_o}[1 + m \cos 2\pi F(t - \tau)]$, where m is the modulation index, τ is the time lag due to transmission, and F is the modulation frequency. The range of F is determined by the spectrum of the temporal envelope of speech. The function $m(F)$ is defined as the MTF.

An attractive feature of the MTF is that the effects of reverberation, ambient noise and the contribution of direct field, which are usually treated individually, are combined in a natural way in the single function $m(F)$. To evaluate the effect of a room on speech intelligibility in a simple manner, the MTF is converted into a single index [1.40,1.41], the STI

$$STI = \sum_{f=125\,\mathrm{Hz}}^{8\,\mathrm{kHz}} w_f \, STI_f \qquad (1.20)$$

where w_f is the frequency weighting factor. From 125 Hz to 8 kHz (octave) w_f is 0·13, 0·14, 0·11, 0·12, 0·19, 0·17 and 0·14, respectively. STI_f can be calculated by

$$STI_f = \frac{[(S/\bar{N})_{\mathrm{app}} + 15]}{30} \qquad (1.21)$$

$$(S/\bar{N})_{\mathrm{app}} = \frac{1}{14} \sum_{F=0\cdot63}^{12\cdot5} (S/N)_{\mathrm{app},F} \qquad (1.22)$$

$$(S/N)_{\mathrm{app},F} = 10\log\frac{m(F)}{1 - m(F)} \qquad (1.23)$$

where $(S/N)_{\mathrm{app}}$ is the apparent S/N ratio, and F is in one-third octave intervals. $(S/N)_{\mathrm{app},F}$ is clipped between ± 15 dB, namely, $(S/N)_{\mathrm{app},F} = 15$ dB if $(S/N)_{\mathrm{app},F} > 15$ dB, and $(S/N)_{\mathrm{app},F} = -15$ dB if $(S/N)_{\mathrm{app},F} < -15$ dB. $m(F)$ is calculated by

$$m(F) = \frac{\left| \int_0^{\infty} e^{2\pi j F t} r(t)\,\mathrm{d}t \right|}{\int_0^{\infty} r(t)\,\mathrm{d}t} [1 + 10^{(-S/N)/10}]^{-1} \qquad (1.24)$$

where $r(t)$ is the squared impulse response. If the decay process is purely exponential, $r(t)$ takes the form

$$r(t) = \mathrm{const.}\, e^{-13\cdot8t/RT} \qquad (1.25)$$

Consequently, equation (1.24) can be simplified to

$$m(F) = \left[1 + \left(2\pi F \frac{RT}{13\cdot8} \right)^2 \right]^{-1/2} [1 + 10^{(-S/N)/10}]^{-1} \qquad (1.26)$$

The rapid speech transmission index (RASTI) is a simplified version of the STI. It was developed for a fast evaluation of speech intelligibility [1.42,1.43]. The RASTI is restricted to only two octave bands, 500 Hz and 2 kHz, and to four or five modulation frequencies in octaves, namely $F = 1, 2, 4, 8$ Hz for 500 Hz and $F = 0\cdot7$, $1\cdot4$, $2\cdot8$, $5\cdot6$, $11\cdot2$ Hz for 2 kHz. The RASTI is calculated by

$$RASTI = \tfrac{4}{9} STI_{500\,\text{Hz}} + \tfrac{5}{9} STI_{2\,\text{kHz}} \tag{1.27}$$

Based on a series of articulation tests, it has been demonstrated that the STI and RASTI are highly correlated with articulation scores [1.39–1.46]. It has also been shown that the early decay of a room governs the MTF and the STI.

1.6. References

1.1. EGAN M. D. *Architectural Acoustics*. McGraw-Hill, New York, 1988.

1.2. LEONARD R. W., DELSASSO L. P. and KNUDSEN V. O. Diffraction of sound by an array of rectangular reflective panels. *Journal of the Acoustical Society of America*, 1964, **36**, 2328–2333.

1.3. TEMPLETON D. (ed.) *Acoustics in the Built Environment*. Butterworth Architecture, Oxford, 1993.

1.4. INSTITUTE OF BUILDING PHYSICS. *Handbook of Architectural Acoustics*. China Building Industry Press, Beijing, 1985 (in Chinese).

1.5. KANG J. and FUCHS H. V. Predicting the absorption of open weave textiles and micro-perforated membranes backed by an airspace. *Journal of Sound and Vibration*, 1999, **220**, 905–920.

1.6. MAA D. Y. Microperforated-panel wideband absorbers. *Noise Control Engineering Journal*, 1987, **29**, 77–84.

1.7. SABINE W. C. *Collected papers on acoustics*. Dover, New York, 1964.

1.8. AMERICAN NATIONAL STANDARDS INSTITUTE (ANSI). *Method for the calculation of absorption of sound by the atmosphere*. ANSI S1.26. ANSI, 1995 (Revised 1999).

1.9. INTERNATIONAL ORGANISATION FOR STANDARDISATION (ISO). *ISO 3382: Acoustics — Measurement of the reverberation time of rooms with reference to other acoustical parameters*. Geneva, 1997.

1.10. BERANEK L. L. *Acoustics*. McGraw-Hill, New York, 1954.

1.11. KROKSTAD A., STRØM S. and SØRSDAL S. Calculating the acoustical room response by use of a ray tracing technique. *Journal of Sound and Vibration*, 1968, **8**, 118–125.

1.12. BORISH J. Extension of the image model to arbitrary polyhedra. *Journal of the Acoustical Society of America*, 1984, **75**, 1827–1836.

1.13. LEE H. and LEE B. An efficient algorithm for image method technology. *Applied Acoustics*, 1988, **24**, 87–115.

1.14. KULOWSKI A. Algorithmic representation of the ray tracing technique. *Applied Acoustics*, 1984, **18**, 449–469.

1.15. LEHNERT H. Systematic errors of the ray-tracing algorithm. *Applied Acoustics*, 1993, **38**, 207–221.

1.16. STEPHENSON U. M. Quantized pyramidal beam tracing — a new algorithm for room acoustics and noise immission prognosis. *Acustica/Acta Acustica*, 1996, **82**, 517–525.

1.17. DRUMM I. A. and LAM Y. M. The adaptive beam-tracing algorithm. *Journal of the Acoustical Society of America*, 2000, **107**, 1405–1412.

1.18. MOORE G. R. *An approach to the analysis of sound in auditoria*. PhD dissertation, University of Cambridge, UK, 1984.

1.19.   LEWERS T. A combined beam tracing and radiant exchange computer model of room acoustics. *Applied Acoustics*, 1993, **38**, 161–178.

1.20.   KANG J. Sound propagation in street canyons: comparison between diffusely and geometrically reflecting boundaries. *Journal of the Acoustical Society of America*, 2000, **107**, 1394–1404.

1.21.   SIEGEL R. and HOWELL J. *Thermal radiation heat transfer*. Hemisphere, Washington, DC, 1981, 2nd edn.

1.22.   SILLION F. X. and PUECH C. *Radiosity and global illumination*. Morgan Kaufmann Publishers, Inc., 1994.

1.23.   FOLEY J. D., VAN DAM A., FEINER S. K. and HUGHES J. F. *Computer graphics: principle and practice*. Addison-Wesley Publishing Company, 1990, 2nd edn.

1.24.   VORLÄNDER M. Simulation of the transient and steady state sound propagation in rooms using a new combined ray tracing/image source algorithm. *Journal of the Acoustical Society of America*, 1989, **86**, 172–178.

1.25.   DALENBÄCK B.-I. L. Room acoustic prediction based on a unified treatment of diffuse and specular reflection. *Journal of the Acoustical Society of America*, 1996, **100**, 899–909.

1.26.   SUH J. S. and NELSON P. A. Measurement of transient response of rooms and comparison with geometrical acoustic models. *Journal of the Acoustical Society of America*, 1999, **105**, 2304–2317.

1.27.   WRIGHT J. R. An exact model of acoustic radiation in enclosed spaces. *Journal of the Audio Engineering Society*, 1995, **43**, 813–819.

1.28.   EASWARAN V. and CRAGGS A. An application of acoustic finite element models to finding the reverberation times of irregular rooms. *Acustica/Acta Acustica*, 1996, **82**, 54–64.

1.29.   HOTHERSALL D. C., HOROSHENKOV K. V. and MERCY S. E. Numerical modelling of the sound field near a tall building with balconies near a road. *Journal of Sound and Vibration*, 1996, **198**, 507–515.

1.30.   SPANDÖCK F. Akustische Modellversuche. *Annalen der Physik*, 1934, **20**, 345–360.

1.31.   BARRON M. Auditorium acoustic modelling now. *Applied Acoustics*, 1983, **16**, 279–290.

1.32.   POLACK J. D., MARSHALL A. H. and DODD G. Digital evaluation of the acoustics of small models: The MIDAS package. *Journal of the Acoustical Society of America*, 1989, **85**, 185–193.

1.33.   ELS H. and BLAUERT J. A measuring system for acoustic scale models. *Proceedings of the Vancouver Symposium on Acoustics and Theatre Planning for the Performing Arts*, 1986, 65–70.

1.34.   CREMER L. and MÜLLER H. A. *Principles and Applications of Room Acoustics, Vol. II*. Applied Science Publishers, Barking, England, 1982.

1.35.   BERANEK L. L. *Acoustic Measurements*. John Wiley and Sons, New York, 1949.

1.36.   AMERICAN NATIONAL STANDARDS INSTITUTE (ANSI). *Method for measuring the intelligibility of speech over communications system*. ANSI S3.2. ANSI, 1989 (Revised 1999).

1.37.   AMERICAN NATIONAL STANDARDS INSTITUTE (ANSI). *Methods for the calculation of the articulation index*. ANSI S3.5. ANSI, 1997.

1.38.   HOUTGAST T. and STEENEKEN H. J. M. The modulation transfer function in room acoustics as a predictor of speech intelligibility. *Acustica*, 1973, **28**, 66–73.

1.39.   HOUTGAST T., STEENEKEN H. J. M. and PLOMP R. Predicting speech intelligibility in rooms from the modulation transfer function I. General room acoustics. *Acustica*, 1980, **46**, 60–72.

1.40. STEENEKEN H. J. M. and HOUTGAST T. A physical method for measuring speech-transmission quality. *Journal of the Acoustical Society of America*, 1980, **67**, 318–326.

1.41. HOUTGAST T. and STEENEKEN H. J. M. The modulation transfer function in room acoustics. *Brüel & Kjær Technical Review*, 1985, **3**, 3–12.

1.42. HOUTGAST T. and STEENEKEN H. J. M. A multi-language evaluation of the RASTI-method for estimating speech intelligibility in auditoria. *Acustica*, 1984, **54**, 185–199.

1.43. STEENEKEN H. J. M. and HOUTGAST T. RASTI: a tool for evaluating auditoria. *Brüel & Kjær Technical Review*, 1985, **3**, 13–39.

1.44. HOUTGAST T. and STEENEKEN H. J. M. A review of the MTF concept in room acoustics and its use for estimating speech intelligibility in auditoria. *Journal of the Acoustical Society of America*, 1985, **77**, 1069–1077.

1.45. NOMURA H., MIYATA H. and HOUTGAST T. Speech intelligibility and modulation transfer function in non-exponential decay fields. *Acustica*, 1989, **69**, 151–155.

1.46. NOMURA H., MIYATA H. and HOUTGAST T. Speech intelligibility enhancement by a linear loudspeaker array in a reverberant field. *Acustica*, 1993, 77, 253–261.

2. Acoustic theories of long spaces

Acoustic theories and computer models for long spaces are presented in this chapter. In Section 2.1 the unsuitability of classic room acoustic theories for long enclosures is discussed. In Section 2.2, by applying the image source method, theoretical formulae and computer models for long spaces with geometrically reflecting boundaries are given. Section 2.3 considers diffusely reflecting boundaries using the radiosity method. Other acoustic theories and models for long spaces are reviewed in Section 2.4. Section 2.5 presents a method to predict the temporal and spatial distribution of train noise in underground stations. In Section 2.6, a practical method is given for predicting acoustic indices in long enclosures, especially the speech intelligibility of multiple loudspeaker PA systems in underground stations. In this book the long space includes both indoor and outdoor spaces, whereas the long enclosure refers to indoor spaces only.

2.1. Unsuitability of classic theory

Although it is known that the sound field in long enclosures is not diffuse as assumed by classic room acoustic theory, in practice classic formulae are often being used as approximations. In this Section the unsuitability of classic theories for the sound distribution and reverberation in long enclosures is briefly analysed [2.1].

2.1.1. Sound distribution

Equation (1.19) is a commonly used classic formula for calculating sound distribution in an enclosure. It is essentially based on the assumption of a diffuse sound field. From the viewpoint of geometrical acoustics, this assumption means that the sound rays are distributed uniformly over all directions of propagation and there is no difference between time average and directional average. According to equation (1.19), the SPL becomes approximately constant beyond the reverberation radius. However, the assumption of a diffuse field is unsuitable for long enclosures.

Consider a rectangular long enclosure with geometrically reflecting boundaries. Using the image source method, the reflection diagram and the SPL distribution at a receiver from all directions are illustrated in Figure 2.1. In the case of infinite length or highly absorbent end walls (see Figure 2.1(a)), it can be seen that the rays are not uniformly distributed in all directions. Moreover, by comparing the two receivers in Figure 2.1(b), it is evident that with the increase of source-receiver distance, the length of the reflected sound path increases systematically and, thus, the SPL decreases continuously along the length. Note that the source-receiver distance below refers to the horizontal distance along the length, except where indicated. In the case of finite long enclosures with highly reflective end walls (see Figure 2.1(c)), the first-order image-source plane plays a more important part than the others. The effect of the other image-source planes is only significant in the area near the end walls. As a result, the SPL still decreases along the length, although less so (or increases slightly) in the area near the end walls. For comparison, a cube is analysed in Figure 2.1(d). It is seen that the rays are distributed more uniformly in all directions, and the difference of sound energy from various image source planes is much less.

It is interesting to note that, contrary to conventional understanding, equation (1.19) is even more inaccurate when the boundaries of a long enclosure are diffusely reflective, although in this case the sound rays at a receiver are distributed more uniformly in all directions. In comparison with geometrically reflecting boundaries, with diffusely reflecting boundaries the average length of the reflected sound path between a given source and receiver is longer and, thus, more energy is absorbed by the medium. Moreover, with diffusely reflecting boundaries the sound rays have more chances of impinging upon the boundaries and, thus, more energy is absorbed there. As a result, the total energy in a long enclosure with diffusely reflecting boundaries is lower than that with geometrically reflecting boundaries, giving an increase in the absolute attenuation, for a given L_W.

Diffusers could also increase the relative attenuation along the length by changing the sound distribution. A simple case is illustrated in Figure 2.2. For the sake of convenience, assume that the reflected energy from the diffuser is distributed equally to receivers 1 and 2, then, with the diffuser, the sound energy attenuation from receiver 1 to 2 is $2E_r$ less than that of the geometrical reflection. In addition, although with the diffuser the increase in sound energy at both receivers is the same (E_r) (see Figure 2.2(b)), the increase of the SPL at receiver 1 is less than the decrease of the SPL at receiver 2 since $E_1 > E_2$. This is useful for increasing the relative SPL attenuation. Note in Figure 2.2 that with the diffuser the sound energy may be increased at receiver 1. This increase is dependent on the absorption of medium and boundaries.

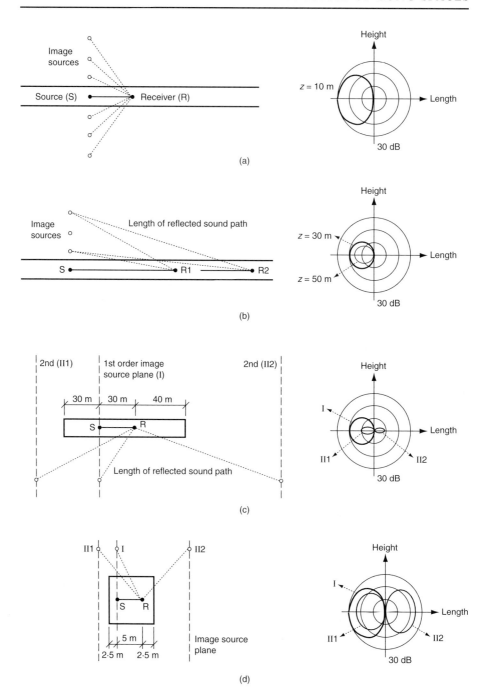

Figure 2.1. Reflection diagram and the directional distribution of SPL in rectangular enclosures with geometrically reflecting boundaries: (a) an infinitely long enclosure; (b) comparison between two receivers; (c) a finite long enclosure; (d) a 10 m cube. The cross-section of the long enclosure is 3 m by 3 m, $\bar{\alpha} = 0.05$

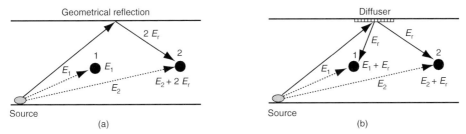

Figure 2.2. Effect of diffusion on the sound attenuation along the length: (a) geometrically reflecting boundaries; (b) with a diffuser

In summary, the above analyses demonstrate the fundamental difference between equation (1.19) and the sound field in long enclosures. That is, instead of becoming stable beyond the reverberation radius, the SPL in long enclosures decreases along the length continuously. As a result, to predict the sound attenuation in long enclosures, classic formulae must not be used.

2.1.2. Reverberation

As mentioned in Chapter 1, the classic Sabine and Eyring formulae are based on the assumption that the sound field is diffuse. Under this assumption the reverberation time is a single value in an enclosure and the decay curves are linear. In the following, the unsuitability of classic reverberation theory for long enclosures is analysed from the viewpoints of both geometrical acoustics and wave theory.

2.1.2.1. Geometrical acoustics

For a rectangular enclosure with geometrically reflecting boundaries, it can be demonstrated that (see Figure 2.1(b)), the difference in sound path length between various sound rays is greater with a shorter source-receiver distance. In other words, the variation in path length of reflected sound decreases systematically along the length. With a greater variation in path length, the reverberation time is shorter but the decay curve is 'sagging' in form (however, if a relatively long time-period is considered, the reverberation time may be longer due to the larger number of reflections with relatively long time-delay). As a result, the reverberation time in long enclosures is fundamentally different from that obtained using the classic Sabine and Eyring formulae: it varies along the length, rather than a single value.

In the case of diffusely reflecting boundaries, Kuttruff has demonstrated that the variation in free path length is greater when the length of a rectangular enclosure becomes longer [2.2]. Consequently, in long enclosures the use of mean free path, which is the basis of classic formulae, becomes unreasonable.

2.1.2.2. *Wave theory*

The eigenfrequencies in a rectangular enclosure with rigid boundaries can be calculated by equation (1.12). The eigenfrequency distributions in a rectangular enclosure with a cross-section of 5 m by 7 m and with three different lengths (9 m, 60 m and 120 m) are shown in Figure 2.3, where $n_x, n_y = 1$–5, $n_z = 1$–10

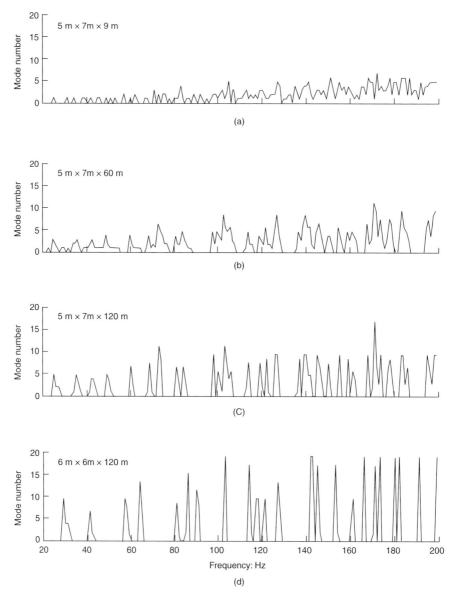

Figure 2.3. Calculated eigenfrequency distribution in rectangular enclosures with various proportions — the vertical axis is the number of modes in a bandwidth of 1 Hz

and the vertical axis is the number of modes in a bandwidth of 1 Hz. It can be seen that when the length is 9 m (see Figure 2.3(a)) the eigenfrequency distribution is rather even. With the increase of the length (see Figure 2.3(b) and 2.3(c)), the eigenfrequency distribution becomes significantly uneven. Furthermore, with approximately the same cross-sectional area, if the cross-section becomes square, 6 m by 6 m, the eigenfrequency distribution is extremely uneven, as shown in Figure 2.3(d).

It can be demonstrated by equation (1.12) that if L is much greater than a and b, the resonant frequency is approximately constant with a wide range of n_z. In other words, for a given frequency, the resonant forms of various modes in a long enclosure are significantly different. Consequently, the difference in damping constants among various modes is likely to be great. Given the fact that the decay curves are linear only when the damping constants are equal for all resonant modes, it appears that in long enclosures the decay curves tend to be non-linear. This might be more significant when a range of frequency rather than a single frequency is considered, since the eigenfrequency distribution at the frequency abscissa is also uneven, as shown in Figure 2.3.

In summary, the above analyses demonstrate that the reverberation characteristics in long enclosures are fundamentally different from those of a diffuse field. Thus, the classic Sabine and Eyring formulae are unsuitable for long enclosures.

2.1.3.   Characterisation of long enclosures

To characterise the 'long enclosure', there appears to be no fixed length/cross-section ratio. Firstly, the absorption condition of end walls should be considered. It has been theoretically and experimentally demonstrated that the increase in reverberation time along the length can be diminished by highly reflective end walls (see Sections 3.2.5 and 5.1). Conversely, with totally absorbent end walls, the reverberation time also varies in a regularly-shaped enclosure (see Section 3.2.6). Secondly, boundary conditions along the length, such as diffusion, could affect the sound field significantly. Thirdly, the width/height ratio should be taken into account. If the width is much greater than the height, an enclosure should be regarded as a flat enclosure rather than a long enclosure, even if the length/cross-section ratio is still great. Finally, the length/cross-section ratio characterising the 'long enclosure' could be different for reverberation and for sound attenuation along the length. As a result, it may be unreasonable to give a fixed length/cross-section ratio to distinguish 'long' and 'regularly-shaped' enclosures. For practical reasons, nevertheless, based on existing measurements and calculations, it is suggested that the long enclosure theory should be applied if the length is greater than six times the width and the height [2.3–2.5].

2.2 Image source method

In many long spaces the boundaries are acoustically smooth and, thus, the image source method can be applied to analyse the sound field. This Section presents theoretical models and formulae based on the method of images. Both long enclosures and urban streets are considered.

2.2.1. Reverberation

2.2.1.1. Image source method I — consideration of every image

The distribution of image sources in an infinitely long enclosure with geometrically reflecting boundaries is shown in Figure 2.4 [2.6]. If a point source is positioned at a distance L_a from the ceiling and L_b from a side wall, and a receiver at a distance z from the source is at the centre of cross-section, the distance between an image source (m, n) and the receiver is

$$D_{z,m,n} = \sqrt{\left[b|m - g(m)| + 2L_b \frac{m}{|m|} g(m)\right]^2 + \left[a|n - g(n)| + 2L_a \frac{n}{|n|} g(n)\right]^2 + z^2}$$

$$(m, n = -\infty \ldots \infty) \qquad (2.1)$$

where

$$g(x) = 1 \qquad \text{for odd } x$$
$$g(x) = 0 \qquad \text{for even } x \qquad (2.2)$$

where a and b are the cross-sectional height and width. If the arrival time of direct sound is defined as $t = 0$, then the reflection of the image source (m, n) arrives at the receiver at time $t_{z,m,n}$

$$t_{z,m,n} = \frac{D_{z,m,n}}{c} - \frac{z}{c} \qquad (2.3)$$

Between time t and $t + \Delta t$ the sound energy contributed by the image source (m, n) at the receiver is

$$E_{z,m,n} = \frac{K_W}{D_{z,m,n}^2} (1 - \alpha)^{|m|+|n|} \qquad \text{if } t \le t_{d,m,n} < t + \Delta t$$

$$= 0 \qquad \qquad \text{otherwise} \qquad (2.4)$$

where K_W is a constant relating to the sound power of the source, and α is the absorption coefficient of the boundaries. It is assumed that all the boundaries have the same absorption coefficient, and the absorption is angle independent.

Between time t and $t + \Delta t$, a short time L_{eq} (equivalent continuous sound level) of the energy responses at the receiver can be calculated by

$$L(t)_z = 10 \log \left[\sum_{m=-\infty}^{\infty} \sum_{n=-\infty}^{\infty} e^{-MD_{z,m,n}} E_{z,m,n} \right] \qquad (2.5)$$

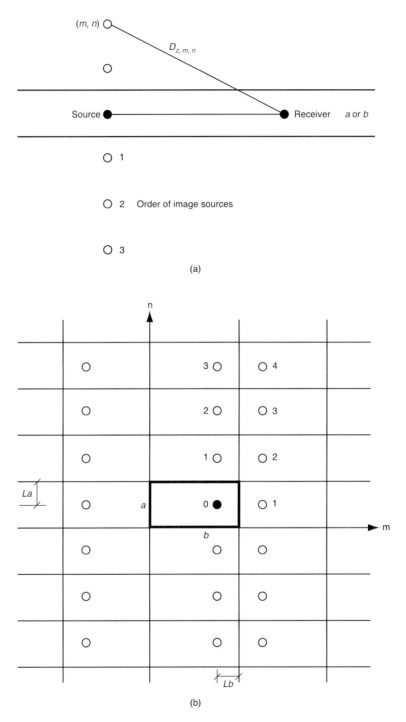

Figure 2.4. Image sources in an infinite rectangular long enclosure: (a) plan or length-wise section; (b) image source plane

where M is the intensity related attenuation constant in air. With equation (2.5) the reverberation time can be derived from the reverse-time integration. Correspondingly, the steady-state SPL at the receiver can be calculated by

$$L_z = 10 \log \sum_{\Delta t} 10^{L(t)_z/10} - L_{\text{ref}} \tag{2.6}$$

where L_{ref} is the reference level.

2.2.1.2.   Image source method II — a statistical method
Infinitely long enclosure

To simplify the above calculation, a statistical method is developed. The basic idea of this method is to calculate (between time t and $t + \Delta t$) the average reflection distance D_0, the approximate number of image sources N, and the average order of image sources R.

When the point source is at the centre of the cross-section, as shown in Figure 2.5, D_0, N and R can be calculated by

$$D_0 = c \left(\frac{z}{c} + t + \frac{1}{2} \Delta t \right) \tag{2.7}$$

$$N = \frac{1}{S} \left\{ \pi \left[c^2 \left(\frac{z}{c} + t + \Delta t \right)^2 - z^2 \right] - \pi \left[c^2 \left(\frac{z}{c} + t \right)^2 - z^2 \right] \right\}$$

$$= \frac{\pi c^2}{S} \left[2 \left(\frac{z}{c} + t \right) + \Delta t \right] \Delta t \tag{2.8}$$

$$R = \frac{2 \Delta \vartheta}{\pi} \sum_{\vartheta = 0, \text{step:} \Delta \vartheta}^{\pi/2} \left(\frac{D_p \sin \vartheta}{a} + \frac{D_p \cos \vartheta}{b} \right) \tag{2.9}$$

In equations (2.7) and (2.8) the term z/c is introduced to scale the arrival time of direct sound to zero. S is the cross-sectional area. In equation (2.9) ϑ is an angle for determining the position of image sources, and D_p is the projection of D_0 into the image source plane

$$D_p = \frac{1}{2} \left[\sqrt{c^2 \left(\frac{z}{c} + t + \Delta t \right)^2 - z^2} + \sqrt{c^2 \left(\frac{z}{c} + t \right)^2 - z^2} \right] \tag{2.10}$$

For $D_p \gg a, b$, $L(t)_z$ can be calculated from

$$L(t)_z = 10 \log \left[N \frac{K_W}{D_0^2} (1 - \alpha)^R \right] - M D_0 \tag{2.11}$$

where M is the air absorption factor in dB/m.

Equation (2.11) shows that the reverberation time in long enclosures depends on the source-receiver distance, which is fundamentally different from that of the diffuse field.

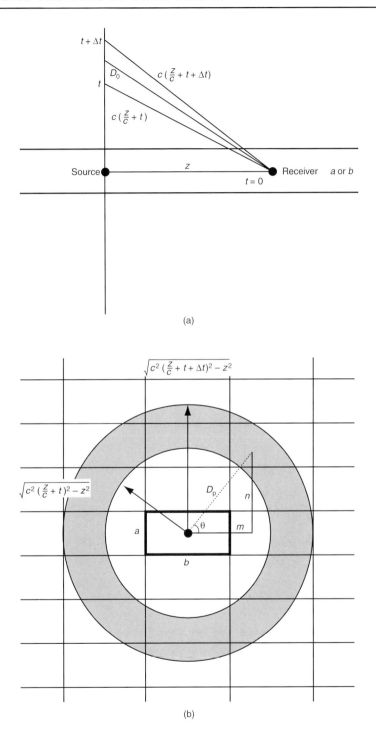

Figure 2.5. Calculation of D_0, N and R between time t and $t + \Delta t$: (a) plan or length-wise section; (b) image source plane

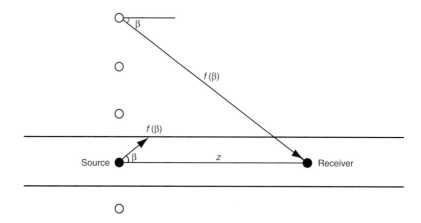

○ Image sources

Figure 2.6. Directional source in an infinite rectangular long enclosure — plan or length-wise section view

Directional source

When the radiation intensity of the source in direction β $(-\pi/2 \le \beta \le \pi/2)$ is $f(\beta)$ $(0 \le f(\beta) \le 1)$, as shown in Figure 2.6, equations (2.5) and (2.11) become

$$L(t)_z = 10\log\left\{f(\beta_{z,m,n})\left[\sum_{m=-\infty}^{\infty}\sum_{n=-\infty}^{\infty}e^{-MD_{z,m,n}}E_{z,m,n}\right]\right\} \tag{2.12}$$

$$L(t)_z = 10\log\left[f(\beta_0)N\frac{K_W}{D_0^2}(1-\alpha)^R\right] - MD_0 \tag{2.13}$$

where

$$\beta_{z,m,n} = \arccos\left(\frac{z}{D_{z,m,n}}\right) \tag{2.14}$$

$$\beta_0 = \arccos\left(\frac{z}{D_0}\right) \tag{2.15}$$

Finite long enclosure

When the end walls are reflective, more image source planes should be considered. As an example, Figure 2.7 shows the distribution of three image source planes. If $2G+1$ image source planes are considered, equation (2.11) becomes

$$L(t)_z = 10\log\sum_{q=-G}^{G}(1-\alpha_e)^{|q|}10^{L(t')_{zq}/10} \tag{2.16}$$

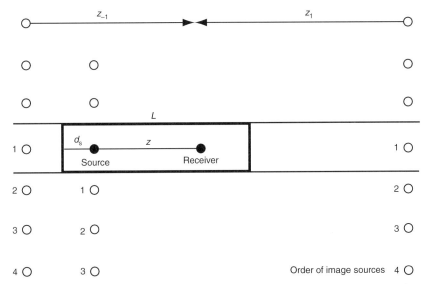

Figure 2.7. Image sources in a finite rectangular long enclosure — plan or length-wise section view

where q is the order of an image source plane, α_e is the absorption coefficient of the end walls, and z_q is the distance between the qth image source plane and the receiver

$$z_q = L|q - g(q)| + 2d_s \frac{q}{|q|} g(q) \qquad (2.17)$$

where d_s is the distance between the source and an end wall, and $g(q)$ is determined in a similar manner to equation (2.2).

In equation (2.16), $L(t')_{z_q}$ is the SPL caused by the qth image source plane between time t and $t + \Delta t$. It can be calculated using equation (2.11) with $z = z_q$ and

$$t' = t - \frac{z_q - z}{c} \geq 0 \qquad (2.18)$$

2.2.1.3. A simplified formula
Providing $\Delta t \ll t$, equations (2.7), (2.8) and (2.10) can be simplified to

$$D_0 \approx z + ct \qquad (2.19)$$

$$N \approx \frac{2\pi c}{S}(z + ct)\Delta t \qquad (2.20)$$

$$D_p \approx \sqrt{ct(2z + ct)} \qquad (2.21)$$

Equation (2.9) can also be approximately expressed by

$$R \approx 0.62 D_{\mathrm{p}} \left(\frac{1}{a} + \frac{1}{b} \right) \tag{2.22}$$

If the decay process is close to linear, the reverberation time at a receiver with a distance of z from the source can, approximately, be calculated by

$$RT_z = \frac{60 t_0}{L(0)_z - L(t_0)_z} \tag{2.23}$$

where t_0 is a coefficient introduced to determine the decay slope. For a linear decay curve RT_z is independent of t_0. Since the decay curves are normally not linear, the approximation by equations (2.23) would be better if t_0 is closer to the actual reverberation time. By substituting equations (2.11) and (2.19) to (2.22) into equation (2.23), the calculation of reverberation time becomes

$$RT_z = \frac{60 t_0}{c t_0 \, \mathrm{M} - 10 \log \left[\frac{z}{z + c t_0} (1-\alpha)^{0.62[(1/a)+(1/b)]\sqrt{c t_0 (2z + c t_0)}} \right]} \tag{2.24}$$

With $t_0 = 2.5\,\mathrm{s}$, which is a typical reverberation time of many underground stations, equation (2.24) becomes

$$RT_z = \frac{150}{850 \, \mathrm{M} - 10 \log \left[\frac{z}{z + 850} (1-\alpha)^{25.6[(1/a)+(1/b)]\sqrt{z + 425}} \right]} \tag{2.25}$$

2.2.1.4.   Comparison between the three methods

In the above Sections three methods of calculating reverberation in long enclosures are given, namely equations (2.5), (2.11) and (2.24). A comparison of $L(t)$ curves between equations (2.5) and (2.11), namely the image source methods I and II, is shown in Figure 2.8, where $S = 6\,\mathrm{m}$ by $4\,\mathrm{m}$, $z = 30\,\mathrm{m}$, $\alpha = 0.1$, $M = \mathrm{M} = 0$, and $\Delta t = 5\,\mathrm{ms}$. The source is at the centre of cross-section. It can be seen that the curve calculated by equation (2.5) is jagged in form. This is caused by some image sources with relatively low order but long reflection distance. For example, for image source (0,2) the reflection distance is $12\,\mathrm{m}$, which is longer than that of image source (1,2), $7.2\,\mathrm{m}$. Using equation (2.11) the peaks and troughs are averaged.

A comparison between equation (2.11) and its simplified version, equation (2.24), is shown in Figure 2.9, where $S = 6\,\mathrm{m}$ by $4\,\mathrm{m}$, $M = 0$, and $\Delta t = 5\,\mathrm{ms}$. In the calculation with equation (2.24), t_0 is $0.9\,\mathrm{s}$ for $\alpha = 0.1$ and $0.4\,\mathrm{s}$ for $\alpha = 0.2$, which are based on the average of RT30 and EDT obtained using equation (2.11). It can be seen that the calculation by equation (2.24) is close to the RT30 given by equation (2.11).

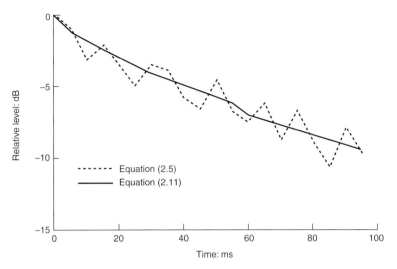

Figure 2.8. Comparison of L(t) curves between equation (2.5) and equation (2.11) — S = 6 m by 4 m, z = 30 m, α = 0·1 and M = M = 0

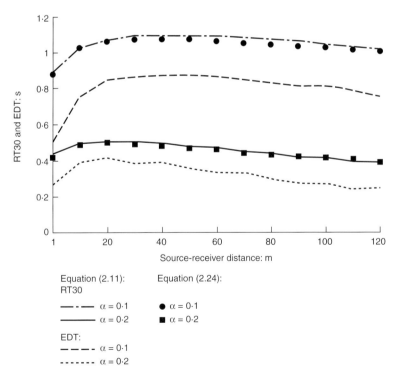

Figure 2.9. Comparison between equation (2.11) and its simplified version, equation (2.24) — S = 6 m by 4 m and M = 0

2.2.1.5. Validation

To validate the above theory, comparison was made between calculation and measurement in a corridor and in several underground stations in London. The calculation was based on the statistical method, namely equations (2.11), (2.13) and (2.16).

A corridor

The length, width and height of the corridor (Cambridge University Engineering Department) were 42·5 m, 1·56 m and 2·83 m, respectively. The boundaries of this corridor could be considered as geometrically reflective. A value $\alpha = 0\cdot1$ appeared to be an appropriate estimation of the average absorption coefficient of the boundaries at 500–1000 Hz. On either end wall there was a door that had an absorption coefficient of about 0·5 at 500–1000 Hz. In the measurement, the sound source — a Brüel & Kjær Sound Source HP1001 — was positioned at the centre of the cross-section and with a distance of 1·5 m from an end wall. The receivers were also along the centre of the cross-section.

A comparison of RT30 between calculation and measurement at 500 Hz and 1000 Hz (octave) is shown in Figure 2.10. The calculation is carried out using equation (2.16), by considering ten image source planes. It is seen that the agreement is very good — the difference in RT30 at the six measurement points is generally within ±10%. Moreover, in correspondence with the

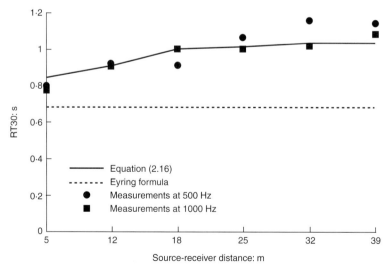

Figure 2.10. Comparison of RT30 between calculation and measurement at 500 Hz and 1000 Hz in the corridor — S = 1·56 m by 2.83 m, L = 42·5 m, $\alpha = 0\cdot1$, $\alpha_e = 0\cdot5$ and M = 0

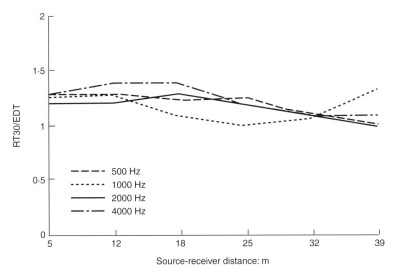

Figure 2.11. Measured RT30/EDT ratios in the corridor

theoretical result, the measured RT30 increases with increasing source-receiver distance. Furthermore, the RT30/EDT ratios are all greater than 1, as shown in Figure 2.11, which means that the decay curves are not linear.

The reverberation time calculated by the Eyring formula is also shown in Figure 2.10. It is evident that the classic theory is far from accurate in long enclosures.

Underground stations

Measurements in London Underground stations were made in two categories [2.7]:

(a)   a cut and cover station of rectangular cross-sectional shape — Euston Square; and

(b)   three deep tube stations of circular cross-sectional shape — Old Street, Warren Street and St John's Wood.

The deep tube stations represented both high and low absorbent conditions — in Old Street station there was an absorbent ceiling and in the other two stations boundaries were acoustically hard.

The length, width and height of Euston Square station were 120 m, 12·9 m and 5·8–6·9 m, respectively. The source, a dodecahedron loudspeaker, was 0·7 m above the platform, 25 m from an end wall, and 1·8 m from the platform edge. The receivers were 1·5 m above the platform and 2 m from the platform edge. The source-receiver distance was 5–40 m. Figure 2.12 shows the measured EDT at 250–4000 Hz (octave). Corresponding to the theory, the EDT increases along the length.

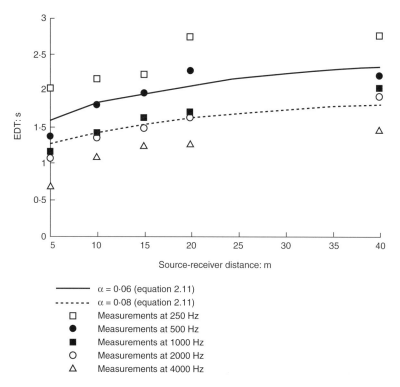

Figure 2.12. Comparison of EDT between the calculation by equation (2.11) and measurements [2.7] in London Euston Square underground station —
$S = 5\cdot85\,m$ *by 13 m and* $M = 0$

The calculation using equation (2.11) is also shown in Figure 2.12. In the calculation α is $0\cdot06$ and $0\cdot08$, which appears to be a reasonable range of the boundary absorption coefficient at 500–1000 Hz. The agreement between calculation and measurement is satisfactory. It is noted, however, that the calculated reverberation time is sensitive to α. In other words, a reasonable estimation of the boundary absorption coefficient is vital for a good prediction. Moreover, the ceiling diffusion may cause differences between calculation and measurement.

For deep tube stations the comparison between measurements and the theory can only be made qualitatively since the formulae are based on rectangular long enclosures. Figure 2.13 shows the measured RT30 and EDT (average of 500 Hz and 1000 Hz) in the three deep tube stations. In the measurements the source was 10 m from an end wall and 2 m from the vertical tangent. The receiver positions were similar to those of Euston Square station. Corresponding to the theory, as the source-receiver distance increases, the RT30 and EDT increase along the length until about 40 m, and then become approximately stable or decrease slightly (see Section 3.2.1). Again, the

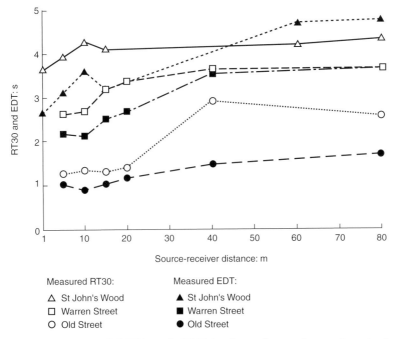

Figure 2.13. Measured RT30 and EDT in three deep tube stations in London [2.7] — average of 500 Hz and 1000 Hz

RT30 is generally greater than the EDT, which indicates that the decay curves are not linear.

The variation in reverberation along the length has also been demonstrated by a series of scale model tests of underground stations, as described in Chapter 5. Figure 2.14 shows two decay curves at 630 Hz (octave) measured in a 1:16 scale model of St John's Wood station. The source-receiver distances are 6 m and 50 m, respectively. It is evident that the decay curves are not linear. At 6 m the decay is more curved than that at 50 m, which suggests that the distribution of reflected sound path lengths varies along the length.

2.2.2. Sound distribution
In the above Section particular attention is paid to reverberation, although the steady-state sound distribution can be calculated using equation (2.6). This Section reviews formulae based on the image source method for calculating sound distribution in long enclosures [2.8].

2.2.2.1. Point source
The sound energy density in an infinite rectangular long enclosure can be derived easily using the image source method. The simplest case is that all the boundaries have the same absorption coefficient, and a point source and

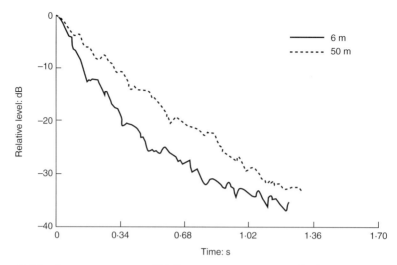

Figure 2.14. Decay curves at 630 Hz at two receivers which are 6 m and 50 m from the source — measured in a 1 : 16 scale model of London St John's Wood station

a receiver are positioned at the centre of the cross-section [2.9]

$$u = \frac{P}{4\pi c} \sum_{m=-\infty}^{\infty} \sum_{n=-\infty}^{\infty} \frac{\rho^{|m|+|n|}}{(ma)^2 + (nb)^2 + z^2} \tag{2.26}$$

where u is the sound energy density, P is the sound power of the source, m and n are the image source orders, and ρ is the mean reflection coefficient of the boundaries. Equation (2.26) is similar to equation (2.4).

An approximation was proposed by Kuttruff [2.9] as

$$u \approx \frac{P}{4\pi c z^2}\left[1 + \frac{4\rho}{(1-\rho)^2}\right] \qquad (z \gg a, b) \tag{2.27}$$

Kuno *et al.* also developed a method to simplify the summation in equation (2.26) [2.10]. The sound energy density at a receiver is

$$u = -\frac{P}{2S}[\cos a_2 z \, ci(a_2 z) + \sin a_2 z \, si(a_2 z)] \tag{2.28}$$

where $a_2 = (U/\pi S)\log\rho$, U is the cross-sectional perimeter, and S is the cross-sectional area.

Computations showed that the results of equations (2.26) and (2.28) were very similar [2.10]. The differences of the two equations were within ± 1 dB in the near field and became less with the increase of source-receiver distance. Equation (2.28) was validated by the measurements in a corridor of 2 m by 2.6 m by 57 m with an accuracy of around 1·5 dB. The measurement frequencies were from 500 Hz to 4 kHz in octaves. The measurements, however, were limited within a relatively low boundary absorption, 0·04–0·07.

Both equations (2.26) and (2.28) are unable to consider the difference of absorption coefficient among various boundaries and, thus, the results could be inaccurate when this difference is great, such as a hard-walled corridor with a strongly absorbent ceiling. Yamamoto proposed a formula to consider the difference of absorption among the four boundaries in infinite rectangular long enclosures [2.11], that was

$$
\begin{aligned}
L_z = L_W - 11 - 10 \log \Bigg[& \frac{1}{z^2} + \sum_{m=0}^{\infty} \sum_{n=0}^{\infty} \frac{(\rho_f \rho_c)^n (\rho_{w1} \rho_{w2})^m}{z^2 + (2n+1)^2 a^2 + (2m+1)^2 b^2} \\
& \times (\rho_c \rho_{w1} + \rho_f \rho_{w1} + \rho_c \rho_{w2} + \rho_f \rho_{w2}) \\
& + \sum_{m=0}^{\infty} \sum_{n=1}^{\infty} 2 \frac{(\rho_f \rho_c)^n (\rho_{w1} \rho_{w2})^m}{z^2 + (2n)^2 a^2 + (2m+1)^2 b^2} (\rho_{w1} + \rho_{w2}) \\
& + \sum_{m=1}^{\infty} \sum_{n=0}^{\infty} 2 \frac{(\rho_f \rho_c)^n (\rho_{w1} \rho_{w2})^m}{z^2 + (2n+1)^2 a^2 + (2m)^2 b^2} (\rho_f + \rho_c) \\
& + \sum_{m=1}^{\infty} \sum_{n=1}^{\infty} 4 \frac{(\rho_f \rho_c)^n (\rho_{w1} \rho_{w2})^m}{z^2 + (2n)^2 a^2 + (2m)^2 b^2} \\
& + \sum_{n=1}^{\infty} 2 \frac{(\rho_f \rho_c)^n}{z^2 + (2n)^2 a^2} + \sum_{n=0}^{\infty} \frac{(\rho_f \rho_c)^n}{z^2 + (2n+1)^2 a^2} (\rho_f + \rho_c) \\
& + \sum_{m=1}^{\infty} 2 \frac{(\rho_{w1} \rho_{w2})^m}{z^2 + (2m)^2 b^2} + \sum_{m=0}^{\infty} \frac{(\rho_{w1} \rho_{w2})^m}{z^2 + (2m+1)^2 b^2} (\rho_{w1} + \rho_{w2}) \Bigg]
\end{aligned}
\quad (2.29)
$$

where ρ_c, ρ_f, ρ_{w1}, ρ_{w2} are the reflection coefficients of the ceiling, floor and two side walls, and the source and receiver are still at the centre of the cross-section. The results of equations (2.26) and (2.29) will be the same if all the boundaries have a uniform absorption coefficient.

The calculations using equation (2.29) showed good agreement with the measurements in a corridor of cross-section 1·76 m by 2·74 m. The absorption coefficients of the ceiling, floor and side walls were $\alpha_c = 0·34$–$0·63$, $\alpha_f = 0·03$, and $\alpha_w = 0·31$–$0·74$, respectively. The accuracy was within about ± 2 dB from 150 Hz to 4800 Hz. This indicated that equation (2.29) could be valid down to low frequencies. Moreover, the results suggested that equation (2.29) could give an effective prediction even if the boundary absorption was relatively high. Unfortunately, the measurements were made with a maximum source-receiver distance of 18 m and, thus, the conclusion was limited.

Redmore used a computer model with the ray image theory to predict the sound attenuation along rectangular corridors [2.12]. Two absorption co-efficients, namely the average of ceiling and floor, and the average of two side walls, were considered. This model, like equations (2.26) and (2.29), gave good agreement with both site and scale model measurements in hard-walled corridors. However, it tended to overestimate the sound levels for corridors containing more highly absorbent material on the floor and ceiling as the distance from the source increased. This could be an important complement of Yamamoto's above conclusion, as in their measurements the cross-section and absorption materials (porous absorbers) were similar but the corridor Redmore used was around 20 m longer. A possible explanation for this overestimation could be that for highly absorbent boundaries the exclusion of diffusion and angle-dependent absorption was unreasonable. In other words, the assumption of geometrical reflection might be inapplicable in this case.

The angle dependence of the absorption coefficient was considered by Sergeev when he derived a series of formulae for long enclosures in a similar manner to the above [2.13,2.14]. This consideration, however, was limited to relatively hard boundaries

$$\rho_\theta = e^{-\varsigma/\theta} \qquad (\pi/3 \le \theta \le \pi/2)$$
$$= \text{const} \qquad (0 \le \theta \le \pi/3) \tag{2.30}$$

where θ is the angle between the boundary normal and an incident ray, ρ_θ is the corresponding reflection coefficient. For a boundary with an infinite thickness, ς can be determined by

$$\varsigma = 4\,\text{Re}(1/J) \tag{2.31}$$

For a boundary with a finite thickness

$$\varsigma = -4|J|^2 \text{Im}[J\,ctg(2\pi\sqrt{N^2 - 1}\,d_b/\lambda)] \tag{2.32}$$

where

$$J = \frac{H}{\sqrt{N^2 - 1}} \tag{2.33}$$

$$H = \frac{\rho_b}{\rho_0} \tag{2.34}$$

$$N = \frac{c}{c_b} \tag{2.35}$$

where λ is the wavelength, d_b is the boundary thickness, ρ_b is the density of the boundary, ρ_0 is the density of air, and c_b is the sound speed in the boundary. Unfortunately, no experimental results were given to demonstrate that the above consideration of the angle-dependent absorption could enable better predictions.

2.2.2.2. Line source

Corresponding to the above investigations on a single source, line sources have also been considered. Similar to equation (2.26), Kuttruff gave a theoretical formula to calculate the sound energy density in infinitely long enclosures with a line source across the width and at half the height of the cross-section [2.9], where the receiver was at the centre of the cross-section

$$u = \frac{P}{4c} \sum_{n=-\infty}^{\infty} \frac{\rho^{|n|}}{\sqrt{(na)^2 + z^2}} \tag{2.36}$$

$$u \approx \frac{P}{4cz} \left(1 + \frac{2\rho}{1 - \rho} \right) \qquad (z \gg a) \tag{2.37}$$

Said derived a formula to consider the different absorption coefficients for the floor, ceiling and side walls in rectangular road tunnels with a line source along the centre of the cross-section and with the same length as the tunnel [2.15,2.16]

$$E_R = \frac{P'}{4a_1 c} \sum_{g,h,q} \frac{(1 - \alpha_c)^g (1 - \alpha_f)^h (1 - \alpha_w)^q a_1}{\pi a_{ghq}}$$

$$\times \left(\arctan \frac{x_0}{a_{ghq}} + \arctan \frac{L - x_0}{a_{ghq}} \right) \tag{2.38}$$

where E_R is the sound energy density from image sources, P' is the sound power per unit length of the line source, g, h and q are the orders of image sources on the ceiling, floor and side walls, $a_1 = 1$ m is the reference distance, a_{ghq} is the distance between the receiver and the image source (g, h, q), x_0 is the horizontal distance between the receiver and a tunnel end, and L is the tunnel length.

To simplify equation (2.38), Said developed a statistical method, where the average absorption coefficient of all the boundaries, $\bar{\alpha}$, was used

$$E_R = \frac{-P'\pi}{2cU \ln(1 - \bar{\alpha})} [2 - e^{[\ln(1 - \bar{\alpha})\pi S x_0/U]} - e^{[\ln(1 - \bar{\alpha})\pi S(L - x_0)/U]}] \tag{2.39}$$

It was demonstrated that equation (2.39) was very close to equation (2.38) when $\bar{\alpha} < 0.4$. In addition, calculations in a tunnel (height, 5 m; width, 15 m; length, 100–500 m) showed that the sound absorption materials on the ceiling were very effective for noise reduction. With an absorbent ceiling the effect of wall absorption was not significant. Unfortunately, in Said's papers no measurement results were given to validate the theory.

2.2.3. Street canyon

An urban street with buildings along both sides can be approximately regarded as a long enclosure with a totally absorbent ceiling. Figure 2.15 illustrates the distribution of image sources in an idealised street, where the street height is a, the street width is b, and a point source S is at (S_x, S_y, S_z) [2.17]. The façades

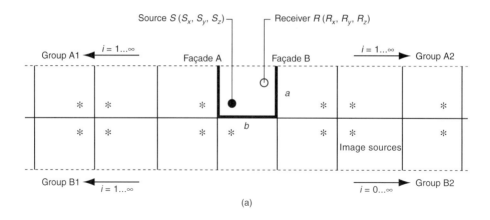

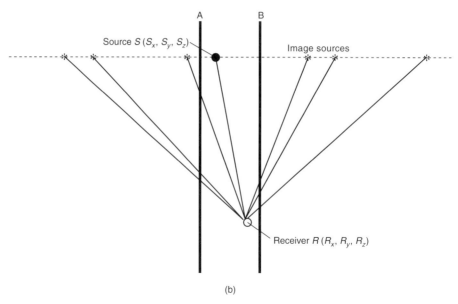

Figure 2.15. Distribution of image sources in an idealised street canyon: (a) cross-section; (b) plan view

are called A and B, and the ground is called G. By comparing Figures 2.4 and 2.15, it can be seen that due to the 'ceiling' absorption, there are only two lines of image source in the street canyon. For calculation convenience, the image sources are divided into four groups, namely A1, A2, B1 and B2. Groups A1 and A2 correspond to the reflections between two façades, and groups B1 and B2 include the reflection from the street ground. With reference to Figure 2.15, the energy from an image source to a receiver R at (R_x, R_y, R_z) can be determined easily. First, consider an image source i $(i = 1 \ldots \infty)$ in group A1. For odd values of i the energy to the receiver is

$$E_i(t) = \frac{1}{4\pi d_i^2}(1 - \alpha_A)^{(i+1)/2}(1 - \alpha_B)^{(i-1)/2}e^{-Md_i} \qquad \left(t = \frac{d_i}{c}\right) \qquad (2.40)$$

where α_A and α_B are the absorption coefficient of façades A and B, respectively. d_i is the distance from the image source i to the receiver

$$d_i^2 = (S_x - R_x)^2 + [(i-1)b + S_y + R_y]^2 + (S_z - R_z)^2 \qquad (2.41)$$

For even i

$$E_i(t) = \frac{1}{4\pi d_i^2}(1 - \alpha_A)^{i/2}(1 - \alpha_B)^{i/2}e^{-Md_i} \qquad \left(t = \frac{d_i}{c}\right) \qquad (2.42)$$

with

$$d_i^2 = (S_x - R_x)^2 + (ib - S_y + R_y)^2 + (S_z - R_z)^2 \qquad (2.43)$$

For an image source i ($i = 1 \ldots \infty$) in group A2, with odd values of i, the sound energy to the receiver is

$$E_i(t) = \frac{1}{4\pi d_i^2}(1 - \alpha_A)^{(i-1)/2}(1 - \alpha_B)^{(i+1)/2}e^{-Md_i} \qquad \left(t = \frac{d_i}{c}\right) \qquad (2.44)$$

with

$$d_i^2 = (S_x - R_x)^2 + [(i+1)b - S_y - R_y]^2 + (S_z - R_z)^2 \qquad (2.45)$$

For even i

$$E_i(t) = \frac{1}{4\pi d_i^2}(1 - \alpha_A)^{i/2}(1 - \alpha_B)^{i/2}e^{-Md_i} \qquad \left(t = \frac{d_i}{c}\right) \qquad (2.46)$$

with

$$d_i^2 = (S_x - R_x)^2 + (ib + S_y - R_y)^2 + (S_z - R_z)^2 \qquad (2.47)$$

For groups B1 and B2, the energy from the image sources to the receiver can be determined using equations (2.40) to (2.47) but replacing the term $S_z - R_z$ with $S_z + R_z$ and also, considering the ground absorption α_G. By summing the energy from all the image sources in groups A1, A2, B1 and B2, and taking direct sound transfer into account, the energy response at the receiver can be obtained. Consequently, the acoustic indices such as the EDT, RT30 and steady-state SPL, can be determined [2.17].

Using the image source method, models have been developed for urban streets with various configurations [2.18]. Consideration has also been given to the interference effect due to multiple reflections [2.19]. This is especially useful for relatively low frequencies and narrowband noise.

2.3. Radiosity method

The theories presented in Section 2.2 are applicable to long spaces with acoustically smooth boundaries. If there are irregularities on boundaries, diffuse reflections should be taken into account. This Section presents a theoretical/ computer model based on the radiosity method [2.20,2.21]. It is assumed that the sound energy reflected from a boundary is dispersed over all directions according to the Lambert cosine law, namely, $I(\eta_1) \sim \cos \eta_1$, where η_1 is the angle between the boundary normal and a reflection.

The general principles of the radiosity method are described in Section 1.4.1. The main steps of this model are:

(a) divide each boundary into certain amount of patches;
(b) distribute the sound energy of an impulse source to the patches — the patches can then be regarded as sound sources, which are called first-order patch sources below;
(c) determine the form factors between pairs of patches;
(d) re-distribute the sound energy of each first-order patch source to every other patch and, thus, generate the second-order patch sources — continue this process and the kth order patch sources can be obtained ($k = 1 \ldots \infty$); this process is 'memory-less', that is, the energy exchange between patches depends only on the form factors and the patch sources of the preceding order; and
(e) calculate the energy response at each receiver by considering all orders of patch sources, from which the acoustic indices can be derived.

2.3.1. Patches

Consider a rectangular long enclosure with a length of L, height of a and width of b, as shown in Figure 2.16, where the boundaries are defined as: C, ceiling; F, floor; A, side wall at $y = 0$; B, side wall at $y = b$; U, end wall at $x = 0$; and V, end wall at $x = L$. Also define the patches along the length, width and height as l ($l = 1 \ldots N_X$), m ($m = 1 \ldots N_Y$) and n ($n = 1 \ldots N_Z$), respectively.

The finer the patch parameterisation, the better the results. However, there is a square-law increase of calculation time in the patch number. To reduce the number of patches, the boundaries are so divided that a patch is smaller when it is closer to an edge. This is because, for a given patch size, the calculation of form factor becomes less accurate the closer the patch is to an edge [2.22]. For the convenience of computation, the division of boundaries is in a manner of geometrical series. Along the width and height, the patch sizes dd_m and dd_n increase from the edges to the centre of a boundary, that is, for example

$$dd_m = k_y q_y^{m-1} \qquad \left(1 \le m \le \frac{N_Y}{2}\right)$$

$$= k_y q_y^{N_Y - m} \qquad \left(\frac{N_Y}{2} < m \le N_Y\right) \qquad\qquad (2.48)$$

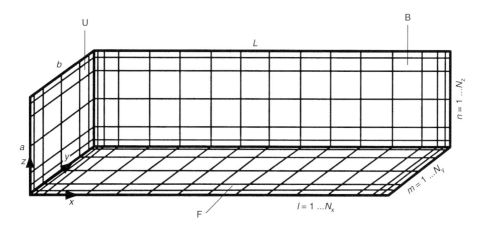

Figure 2.16. Three-dimensional projection of an idealised long enclosure showing an example division of the boundaries with $N_X = 16$, $N_Y = N_Z = 6$ and $q_x = q_y = q_z = 2$

where q_y ($q_y > 1$) is the ratio between two adjacent patches, N_Y should be an even number, and

$$k_y = \frac{b}{2}\frac{1 - q_y}{1 - q_y^{N_Y/2}} \qquad (2.49)$$

Along the length, to avoid too great differences between patches, the patch size dd_l increases from $l = 1$ to $N_X/4$, decreases from $l = 3N_X/4 + 1$ to N_X, and is constant between $l = N_X/4 + 1$ and $3N_X/4$

$$dd_l = k_x q_x^{l-1} \qquad \left(1 \le l \le \frac{N_X}{4}\right)$$

$$= k_x q_x^{(N_X/4)-1} \qquad \left(\frac{N_X}{4} < l \le \frac{3N_X}{4}\right)$$

$$= k_x q_x^{N_X - l} \qquad \left(\frac{3N_X}{4} < l \le N_X\right) \qquad (2.50)$$

where q_x ($q_x > 1$) is the ratio between two adjacent patches, N_X should be divisible by 4, and

$$k_x = \frac{L}{2}\frac{1}{\dfrac{1 - q_x^{(N_X/4)}}{1 - q_x} + \dfrac{N_X}{4}q_x^{(N_X/4)-1}} \qquad (2.51)$$

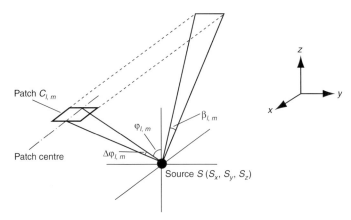

Figure 2.17. Distribution of the energy from a point source to a patch on the ceiling

Correspondingly, the coordinates of the centre of a patch, for example, $C_{l,m}$, can be determined by

$$d_l = -\frac{1}{2}dd_l + \sum_{l=1}^{l} dd_l \quad \text{and} \quad d_m = -\frac{1}{2}dd_m + \sum_{m=1}^{m} dd_m \qquad (2.52)$$

As an example, the division of boundaries with $N_X = 16$, $N_Y = N_Z = 6$ and $q_x = q_y = q_z = 2$ is shown in Figure 2.16.

2.3.2. Distribution of source energy to patches

Consider a point source S at (S_x, S_y, S_z) generating an impulse. The fraction of the energy received by each patch is the same as the ratio of the solid angle subtended by the patch at the source to the total solid angle. A first-order patch source, $C_1(t)_{l,m}$, for example (see Figure 2.17), can be calculated by

$$C_1(t)_{l,m} = 0 \quad \text{except}$$

$$C_1\left(\frac{S_{l,m}}{c}\right)_{l,m} = K_W(1 - \alpha_{C_{l,m}})e^{-MS_{l,m}}\frac{1}{4\pi}\int_0^{\beta_{l,m}}\int_{\varphi_{l,m}}^{\varphi_{l,m}+\Delta\varphi_{l,m}}\cos\varphi\,d\varphi\,d\beta$$

$$= K_W(1 - \alpha_{C_{l,m}})e^{-MS_{l,m}}\frac{1}{4\pi}\left|\sin(\varphi_{l,m} + \Delta\varphi_{l,m}) - \sin\varphi_{l,m}\right|\beta_{l,m}$$

$$(2.53)$$

where t is the time and $t = 0$ represents the moment at which the source generates an impulse. $S_{l,m}$ is the mean beam length between the source and patch $C_{l,m}$. K_W is a constant relating to the sound power of the source. $\alpha_{C_{l,m}}$ is the angle-independent absorption coefficient of patch $C_{l,m}$. $\varphi_{l,m}$, $\Delta\varphi_{l,m}$ and

$\beta_{l,m}$ are the angles determining the location of patch $C_{l,m}$ with reference to the source. These angles can be calculated by

$$\sin(\varphi_{l,m} + \Delta\varphi_{l,m}) = \frac{d_l + \frac{1}{2}dd_l - S_x}{\sqrt{(d_l + \frac{1}{2}dd_l - S_x)^2 + (d_m - S_y)^2 + (a - S_z)^2}} \qquad (2.54)$$

$$\sin\varphi_{l,m} = k_\varphi \frac{d_l - \frac{1}{2}dd_l - S_x}{\sqrt{(d_l - \frac{1}{2}dd_l - S_x)^2 + (d_m - S_y)^2 + (a - S_z)^2}} \qquad (2.55)$$

and

$$\beta_{l,m} = \left|\arctan\left|\frac{d_m + \frac{1}{2}dd_m - S_y}{a - S_z}\right| - k_\beta \arctan\left|\frac{d_m - \frac{1}{2}dd_m - S_y}{a - S_z}\right|\right| \qquad (2.56)$$

In equations (2.55) and (2.56) k_φ and k_β are used to consider the source position which is between the two sides of patch $C_{l,m}$

$$k_\varphi = -1 \qquad (d_l - \tfrac{1}{2}dd_l \le S_x \le d_l + \tfrac{1}{2}dd_l)$$
$$= 1 \qquad \text{Otherwise} \qquad (2.57)$$

$$k_\beta = -1 \qquad (d_m - \tfrac{1}{2}dd_m \le S_y \le d_m + \tfrac{1}{2}dd_m)$$
$$= 1 \qquad \text{Otherwise} \qquad (2.58)$$

The mean beam length between the source and patch $C_{l,m}$, $S_{l,m}$, can be determined by subdividing patch $C_{l,m}$ into $N_l \times N_m$ $(N_l, N_m \ge 1)$ equal elements and then calculating their average distance to the source

$$S_{l,m} = \frac{1}{N_l N_m} \sum_{i=1}^{N_l} \sum_{j=1}^{N_m} \sqrt{\left[d_l - \frac{1}{2}dd_l + \frac{dd_l}{N_l}\left(i - \frac{1}{2}\right) - S_x\right]^2}$$
$$\overline{+ \left[d_m - \frac{1}{2}dd_m + \frac{dd_m}{N_m}\left(j - \frac{1}{2}\right) - S_y\right]^2 + (a - S_z)^2}$$
$$\qquad (2.59)$$

If the distance between a patch and the source is large relative to the size of the path, the mean beam length can be approximated using the distance from the source to the centre of the patch, namely with $N_l = N_m = 1$ in equation (2.59)

$$S_{l,m} = \sqrt{[d_l - S_x]^2 + [d_m - S_y]^2 + (a - S_z)^2} \qquad (2.60)$$

A directional source can also be modelled. Assume that the radiation strength of a source is $f(\eta_x, \eta_y)$ $(0 \le f(\eta_x, \eta_y) \le 1)$ in the direction of (η_x, η_y), where η_x is the angle between the x axis and the xz plane projection of the line joining the source and a patch centre, and η_y is similarly defined. Then the term $f(\eta_x, \eta_y)$ should be added in equation (2.53). Consider patch $C_{l,m}$, for example, the angles η_x and η_y can be calculated by

$$\eta_x(C_{l,m}) = \arctan\left(\frac{a - S_z}{d_l - S_x}\right) \qquad (2.61)$$

and

$$\eta_y(C_{l,m}) = \arctan\left(\frac{a - S_z}{d_m - S_y}\right) \qquad (2.62)$$

2.3.3. Form factors

In rectangular long enclosures, the relative location between any two patches is either orthogonal or parallel. For orthogonal patches, the form factor can be calculated using Nusselt's method [2.23]. That is, the computing form factor is equivalent to projecting the receiving patch onto a unit hemisphere centred about the radiation patch, projecting this projected area orthographically down onto the hemisphere's unit circle base and dividing by the area of the circle. As an example, Figure 2.18 illustrates the calculation from emitter $A_{l',n'}$ ($l' = 1 \ldots N_X$, $n' = 1 \ldots N_Z$) to receiver $C_{l,m}$. By considering the absorption of patch $A_{l',n'}$ and air absorption, the energy from $A_{l',n'}$ to $C_{l,m}$,

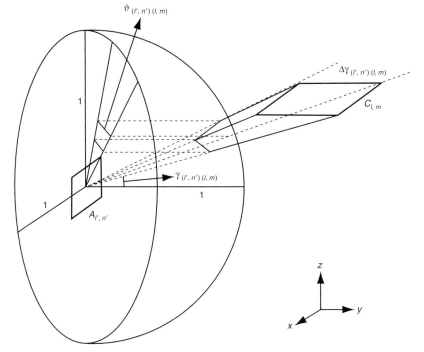

Figure 2.18. Determination of the form factor from emitter $A_{l',n'}$ to an orthogonal patch $C_{l,m}$

$AC_{(l',n'),(l,m)}$, can be calculated by

$$AC_{(l',n'),(l,m)} = (1 - \alpha_{A_{l',n'}})e^{-Md_{(l',n'),(l,m)}} \frac{1}{2\pi}$$

$$\times \left| \cos^2 \gamma_{(l',n'),(l,m)} - \cos^2(\gamma_{(l',n'),(l,m)} + \Delta\gamma_{(l',n'),(l,m)}) \right| \vartheta_{(l',n'),(l,m)}$$

$$(2.63)$$

where $d_{(l',n'),(l,m)}$ is the mean beam length between patches $A_{l',n'}$ and $C_{l,m}$. It can be calculated in a similar manner to equation (2.59), or approximated using the distance between the centres of the two patches

$$d_{(l',n'),(l,m)} = \sqrt{(d_l - d_{l'})^2 + d_m^2 + (a - d_{n'})^2}$$

$$(2.64)$$

$\gamma_{(l',n'),(l,m)}$, $\Delta\gamma_{(l',n'),(l,m)}$ and $\vartheta_{(l',n'),(l,m)}$ are the angles for determining the relative location of the two patches

$$\cos\gamma_{(l',n'),(l,m)} = \frac{d_m - \frac{1}{2}dd_m}{\sqrt{(d_l - d_{l'})^2 + (d_m - \frac{1}{2}dd_m)^2 + (a - d_{n'})^2}}$$

$$(2.65)$$

$$\cos(\gamma_{(l',n'),(l,m)} + \Delta\gamma_{(l',n'),(l,m)}) = \frac{d_m + \frac{1}{2}dd_m}{\sqrt{(d_l - d_{l'})^2 + (d_m + \frac{1}{2}dd_m)^2 + (a - d_{n'})^2}}$$

$$(2.66)$$

$$\vartheta_{(l',n'),(l,m)} = \left| \arctan\left|\frac{d_l - \frac{1}{2}dd_l - d_{l'}}{a - d_{n'}}\right| - k_\vartheta \arctan\left|\frac{d_l + \frac{1}{2}dd_l - d_{l'}}{a - d_{n'}}\right| \right| \quad (2.67)$$

In equation (2.67) k_ϑ is used to consider the case where the two patches have the same coordinate in the length direction

$$k_\vartheta = -1 \qquad (l = l')$$

$$= 1 \qquad \text{Otherwise} \qquad (2.68)$$

For parallel patches, the form factor can be calculated using a method developed by Cohen and Greenberg, that is, projecting the receiving patch onto the upper half of a cube centred about the radiation patch [2.23,2.24]. Consider emitter $F_{l',m'}$ ($l' = 1 \ldots N_X$, $m' = 1 \ldots N_Y$) and receiver $C_{l,m}$ (see Figure 2.19), for example, the energy from $F_{l',m'}$ to $C_{l,m}$, $FC_{(l',m'),(l,m)}$, can be calculated by

$$FC_{(l',m'),(l,m)} = (1 - \alpha_{F_{l',m'}})e^{-Md_{(l',m'),(l,m)}} \frac{a^2 dd_l dd_m}{\pi[(d_l - d_{l'})^2 + (d_m - d_{m'})^2 + a^2]^2}$$

$$(2.69)$$

where $d_{(l',m'),(l,m)}$ is the mean beam length between the two patches. Again, it can be calculated in a similar manner to equation (2.59), or approximated

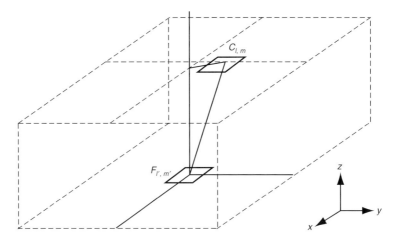

Figure 2.19. Determination of the form factor from emitter $F_{l',m'}$ to a parallel patch $C_{l,m}$

using the distance between the centres of the two patches

$$d_{(l',m'),(l,m)} = \sqrt{(d_l - d_{l'})^2 + (d_m - d_{m'})^2 + a^2} \tag{2.70}$$

2.3.4. Energy exchange between patches

A kth order patch source, such as $C_k(t)_{l,m}$, can be determined by summing the energy contribution from all the boundaries except C

$$C_k(t)_{l,m} = A_{C_k(t)_{l,m}} + B_{C_k(t)_{l,m}} + F_{C_k(t)_{l,m}} + U_{C_k(t)_{l,m}} + V_{C_k(t)_{l,m}} \tag{2.71}$$

where, for example, the contribution from side wall A is $A_{C_k(t)_{l,m}}$, which is the sum of the energy from all the patches on this boundary

$$A_{C_k(t)_{l,m}} = \sum_{l'=1}^{N_X} \sum_{n'=1}^{N_Z} AC_{(l',n'),(l,m)} A_{k-1} \left(t - \frac{d_{(l',n'),(l,m)}}{c} \right)_{l'n'}$$

$$\left(t - \frac{d_{(l'n'),(l,m)}}{c} \geq 0 \right) \tag{2.72}$$

From equation (2.72) it can be seen that the kth order patch sources depend only on the form factors and the $(k-1)$th order patch sources.

2.3.5. Energy at a receiver

At time t the energy at a receiver R (R_x, R_y, R_z) contributed from all the kth order patch sources can be written as

$$E_k(t) = E_k(t)_C + E_k(t)_F + E_k(t)_A + E_k(t)_B + E_k(t)_U + E_k(t)_V \tag{2.73}$$

where, for example, the contribution from the patch sources on boundary C is $E_k(t)_C$, which can be calculated by

$$E_k(t)_C = \sum_{l=1}^{N_X} \sum_{m=1}^{N_Y} \left[\frac{C_k\left(t - \frac{R_{l,m}}{c}\right)_{l,m}}{\pi R_{l,m}^2} \cos(\xi_{l,m}) \right] e^{-MR_{l,m}} \quad \left(t - \frac{R_{l,m}}{c} \geq 0\right)$$

(2.74)

where $R_{l,m}$ is the mean beam length between the receiver and patch $C_{l,m}$. It can be calculated in a similar manner to equation (2.59), or approximated using the distance between the receiver and the patch centre

$$R_{l,m} = \sqrt{(d_l - R_x)^2 + (d_m - R_y)^2 + (a - R_z)^2}$$

(2.75)

$\xi_{l,m}$ is the angle between the normal of patch $C_{l,m}$ and the line joining the receiver and the patch

$$\cos(\xi_{l,m}) = \frac{a - R_z}{\sqrt{(d_l - R_x)^2 + (d_m - R_y)^2 + (a - R_z)^2}}$$

(2.76)

By considering all orders of patch sources as well as the direct energy transport from source to receiver, the energy response at receiver R can be given by

$$L(t) = 10 \log \left[E_d(t) + \sum_{k=1}^{\infty} E_k(t) \right] - L_{\text{ref}}$$

(2.77)

where L_{ref} is the reference level, and the direct energy can be calculated by

$$E_d(t) = \frac{1}{4\pi z^2} e^{-Mz} \qquad \left(t = \frac{z}{c}\right)$$

$$= 0 \qquad\qquad \text{otherwise}$$

(2.78)

where z is the source-receiver distance

$$z = \sqrt{(S_x - R_x)^2 + (S_y - R_y)^2 + (S_z - R_z)^2}$$

(2.79)

By introducing a term z/c to translate the arrival time of direct sound to zero, the decay curve can be obtained by the reverse-time integration of $L(t)$. Consequently, the EDT and RT30 can be derived. The steady-state SPL at receiver R can be calculated by

$$L_R = 10 \log \sum_{\Delta t} 10^{L(t)/10}$$

(2.80)

The above model can be modified to simulate the sound field in street canyons. By reducing the number of boundaries to three, the model has been used for an idealised street canyon as illustrated in Figure 2.15 [2.25,2.26]. In the model the ground can be considered as either diffusely or geometrically reflective. By increasing the boundary number, the model has been extended to consider interconnected streets, such as the configuration illustrated in Figure 4.26 [2.27–2.29]. It is noted that in this case the form factor is zero if there is no line of sight between a pair of patches.

2.3.6. Validation

According to the above theoretical model, a C-program has been developed. The time interval for calculating energy response can be selected, typically 3–5 m/s. Calculation stops when the total energy reduces to a certain amount, typically 10^{-6} of the source energy. There is an initial stage of determining patch division, so that a required accuracy in calculating form factors, typically to three decimal places, can be achieved with a minimum number of patches. The accuracy can be evaluated by the fact that the sum of the form factors from any patch to all the other patches should be unit.

The program was applied to a cube, so as to examine the model's effectiveness and accuracy. It is assumed that the absorption coefficient of the boundaries is uniform. In such an enclosure the sound field should be close to diffuse. This means that the reverberation time should be near the Sabine or Eyring result, the decay curves should be close to linear, and the spatial SPL distribution should be considerably even.

In the calculation the cube side was 10 m, and the absorption coefficient of the boundaries was 0·2. For the sake of convenience, no air absorption was considered, namely $M = 0$. An omnidirectional source was at (3 m, 3 m, 3 m). Receivers were along three diagonals of the cube and, on each diagonal, nine evenly distributed receivers were considered. Each boundary was divided into 144 patches, with $q_x = 1·5$ and $q_y = q_z = 1·3$. In Figure 2.20 the calculation arrangement is illustrated.

Using the above parameters the program calculated the form factors accurate to three decimal places for any patch position on a boundary. Also, if $M = 0$, the sum of the energy in all the first-order patch sources should be equal to the source energy. The program calculated this energy distribution accurate to four decimal places.

The calculated RT30 and EDT in the cube are shown in Figure 2.21. It can be seen that the RT30 is constant at all the receivers, and the RT30 of 1·27 s is close to the Sabine result of 1·34 s and Eyring result of 1·2 s. The EDT is also rather uniform in the cube and close to the Sabine or Eyring result except at the receivers that are very close to the source. Figure 2.22 shows the decay curves at two typical receivers, points 2 and 6 along line 1 (see Figure 2.20). It is seen that the decay curves are generally linear except for some initial fluctuations. Calculation has also shown that the SPL distribution in the cube is considerably even and with a variation of less than 2 dB, except in the very near field.

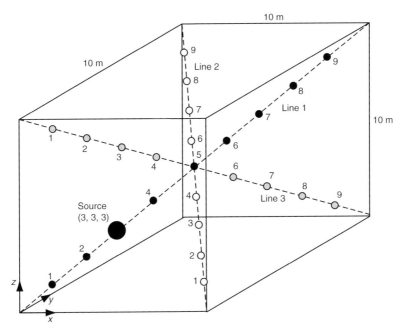

Figure 2.20. Three-dimensional projection of the cube showing the source position and receiver lines

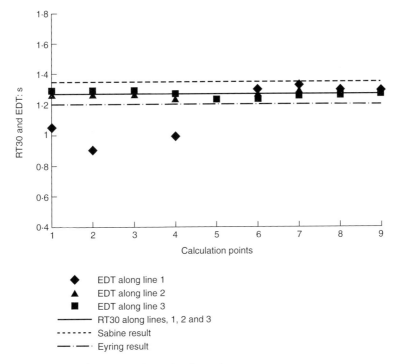

Figure 2.21. Reverberation times in the cube

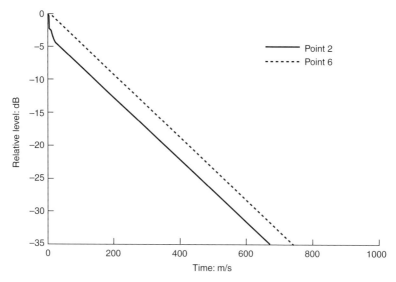

Figure 2.22. Decay curves at point 2 and point 6 along line 1 in the cube

Overall, given the good agreement between the results calculated by the program and the expected acoustic characteristics of the cube according to classic theories, the effectiveness and accuracy of the program is demonstrated. Comparisons were also made between the prediction by the model and the measurements in the scale models of two enclosures [2.20] and a street canyon. Good agreement was obtained.

2.4.   Other theories and models

Following the theoretical/computer models based on the image source method and the radiosity method, this Section reviews some other theories and models relating to acoustics of long spaces.

2.4.1.   Ray tracing and boundary element method

A ray tracing model for long enclosures has been developed by Yang and Shield [2.30,2.31]. In comparison with the image source models, the ray tracing model can consider more complex configurations. Ray tracing models have also been developed for street canyons, such as a model by Thomas for calculating sound distribution in interconnected streets [2.32], and a model by Rahim for considering strategic design options in a single street [2.33].

The boundary element method has been employed for urban streets. For example, a two-dimensional boundary element numerical model was used to study the sound field in the region of balconies in a tall building close to a roadway and it was found that treatment of the ceiling or the rear wall of the balcony is the most efficient in terms of noise reduction [2.34].

2.4.2. Wave theory based on geometrical reflection

In contrast to the specular, single plane wave reflection theory of geometrical acoustics, Davies, in a manner of plane wave decomposition, derived a series of formulae to estimate the sound attenuation along a rectangular corridor [2.35]. The estimates, however, were still based on following the propagation of plane waves on a form of geometrical acoustics. The analyses were at high frequency, at which the modal summations could be replaced by suitable integrals. Attention was limited to the total acoustic power flow and not to the details of the cross-sectional variations of the sound pressure field. The diffraction was ignored. In other words, the uniform impedance condition was assumed. Since the diffraction effects might accumulate after several reflections, only two cases were considered: either highly absorbent materials, where all the energy of a plane wave was absorbed effectively after two reflections, or relatively hard boundary materials, where the true sound field involved only a small perturbation of the rigid boundary case. Two idealised sound sources were assumed. One was the equal energy source with which the propagating modes had equal energy, and the other was a simple source with which there was relatively more energy in the higher order modes.

Based on the above assumptions, a formula for calculating the receiver/input energy ratio was given

$$\frac{P_{SO}}{P_{IN}} = \left(1 + \sum_{i=1}^{4} \frac{P_{ABS,i}}{P_{IN} - P_{ABS,i}}\right)^{-1} \tag{2.81}$$

where $P_{ABS,i} = P_{IN} - P_{SO,i}$. P_{SO} is the power flow at the receiver, P_{IN} is the input power, $i = 1-4$ is the boundary index, and $P_{SO,i}$ is the power flow at the receiver caused by the boundary i. The calculation of $P_{SO,i}$ should be made in four categories, namely, low absorbent boundaries, the simple source and large values of l_3/l_1; low absorbent boundaries, the equal energy source and large values of l_3/l_1; high absorbent boundaries and the simple source; high absorbent boundaries and the equal energy source. Equations (2.82) to (2.85) correspond to the four categories, respectively

$$\frac{P_{SO,i}}{P_{IN}} = \frac{2}{\pi}[ci(y)\sin y - si(y)\cos y] \tag{2.82}$$

$$\frac{P_{SO,i}}{P_{IN}} = -\frac{2}{\pi}[ci(y)(y\cos y - \sin y) + si(y)(y\sin y + \cos y)] \tag{2.83}$$

$$\frac{P_{SO,i}}{P_{IN}} = 1 - 2\alpha_i\left(1 - \frac{2}{\pi}\tan^{-1}\frac{4l_2}{l_3}\right) + \alpha_i^2\left(1 + \frac{2}{\pi}\tan^{-1}\frac{2l_2}{l_3} - \frac{4}{\pi}\tan^{-1}\frac{4l_2}{l_3}\right)$$

$$- \frac{\alpha_i l_3}{4l_i}(1 - \alpha_i)\ln\left(\frac{l_3^2}{16l_2^2 + l_3^2}\right) - \frac{\alpha_i^2 l_3}{4l_i}\ln\left(\frac{l_3^2}{4l_2^2 + l_3^2}\right) \tag{2.84}$$

$$\frac{P_{SO,i}}{P_{IN}} = 1 - 2\alpha_i\left(1 - \frac{2}{\pi}\tan^{-1}\frac{4l_2}{l_3}\right)$$

$$+ \alpha_i^2\left(1 + \frac{2}{\pi}\tan^{-1}\frac{2l_2}{l_3} - \frac{4}{\pi}\tan^{-1}\frac{4l_2}{l_3}\right) \qquad (2.85)$$

where α_i is the absorption coefficient of the boundary i, l_1 is the width of the boundary i, l_2 is the width of the boundary that is vertical to the boundary i, l_3 is the distance between the receiver and the cross-section with P_{IN}, and $y = \alpha_i l_3/2l_2$.

The theoretical estimates were compared to the measurements in two typical corridors in a 1/3 octave band centred at 2000 Hz. The equal energy source was simulated by a loudspeaker placed in a large, hard-walled stairwell at one end of a corridor. The simple source was simulated by a loudspeaker placed at a corner of the other corridor. The agreement was satisfactory but the calculation tended to underestimate the actual attenuation. A possible reason is that the assumed conditions of the theory were too strict to be practically achieved.

2.4.3. An empirical formula

Redmore gave an empirical formula by carrying out a series of measurements in a 1:8 scale model (height, 1·6–3·2 m; width, 1·28–2·48 m; length, 18·4–36·8 m, full scale) [2.36]. The source was a loudspeaker positioned behind a small hole on an end wall. The formulae were as follows

$$L_z = 10\log\rho_0 c + 10\log P + 10\log\left(\frac{1}{2\pi z^2} + \frac{B}{U\bar{\alpha}}10^{-\nabla z/10}\right) - L_{ref} \qquad (2.86)$$

$$L_z = 10\log\left\{\frac{1}{2\pi z^2} + \frac{B}{U\bar{\alpha}}10^{-\nabla z/10} + \rho_E\left[\frac{1}{2\pi(2L-z)^2}\right.\right.$$

$$\left.\left. + \frac{B}{U\bar{\alpha}}10^{-\nabla(2L-z)/10}\right]\right\} \qquad (2.87)$$

where $B = 0·14$ is an empirical attenuation coefficient, $\nabla = 1·4U\bar{\alpha}/S$ is an empirical rate of the sound attenuation along the length outside the direct field, and ρ_E is the reflection coefficient of the end wall opposite the sound source. In equation (2.87) the effect of end walls is taken into account. Equation (2.86) is used for the absolute SPL and equation (2.87) is applied to at least two receivers to predict comparative levels outside the direct field. The mean absolute difference between the predicted and measured values was 0·4 dB in the scale model and 1·4 dB in a corridor.

A comparison between calculations using equations (2.26), (2.82) and (2.87) is shown in Figure 2.23, where $\alpha = 0·05$ and 0·2, $S = 5$ m by 5 m, and the SPL attenuation is with reference to a source-receiver distance of 20 m. In principle,

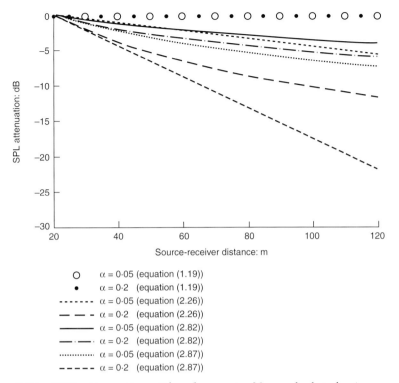

Figure 2.23. SPL attenuation with reference to 20 m calculated using equation (1.19), equation (2.26), equation (2.82) and equation (2.87) — S = 5 m by 5 m

the three formulae are comparative although they are derived using different methods, namely the image source method, wave theory and empirical method. However, it is seen that the attenuation calculated using equation (2.87) is considerably greater than that using equation (2.26). This might be caused by the fact that the calculation range is outside the original range of equation (2.87). Conversely, the attenuation calculated using equation (2.82) is less than that using equation (2.26), especially for $\alpha = 0.2$. This is probably because in equation (2.82) the assumption of 'hard boundary' tends to cause inaccuracy with the increase of absorption coefficient. The calculation by the classic formula, equation (1.19), is also shown in Figure 2.23. As expected, the result is fundamentally different from those by equations (2.26), (2.82) and (2.87).

2.4.4. Consideration of diffusion

Yamamoto [2.11] gave a theoretical formula for the semi-diffuse field, namely, that the sound density was the same in a cross-section

$$L_z = L_{z=0} - 2.17 \frac{\bar{\alpha} U}{S} z \tag{2.88}$$

Comparison with the measurement in a corridor showed that equation (2.88) was far from accurate [2.11]. This seemed to suggest that the assumption of the semi-diffuse field was not applicable, at least in such a corridor.

Ollendorff developed a theoretical method to calculate the noise level and reverberation time in rectangular road tunnels using the partial differential equation of the diffusion of phonons [2.37]. The theory, however, appears too complicated to be used practically.

A series of integral formulae for the sound field resulting from diffusely reflecting boundaries has been developed with similar principles to the radiosity method [2.38–2.41]. However, analytic solutions are only available for limited cases. For long enclosures, formulae were given by Kuttruff for two configurations [2.9]. For a point source at the centre of a circle cross-section

$$u = \frac{P}{4\pi c z^2} + \frac{2\rho P}{\pi^2 a_0^2 c} \int_0^\infty \frac{[\chi(\xi)]^2 \cos\left(\xi \frac{z}{a_0}\right)}{1 - \rho \lambda(\xi)} \, d\xi \tag{2.89}$$

where $\chi(\xi) = \xi K_1(\xi)$, K_1 is the modified Hankel function, a_0 is the radius, and

$$\lambda(\xi) \approx \frac{1}{1 + \frac{4}{3}\xi^2} \tag{2.90}$$

For a line source along the width and in the middle of a rectangular cross-section with geometrically reflecting side walls and diffusely reflecting ceiling and floor, the sound energy density can be calculated by

$$u = \frac{P}{4cz} + \frac{2\rho P}{\pi a c} \int_0^\infty \frac{e^{-\xi} \cos\left(\xi \frac{z}{a}\right)}{1 - \rho \chi(\xi)} \, d\xi \tag{2.91}$$

In equations (2.89) and (2.91) the receivers are along the centre of the cross-section. Using the above equations it has been shown that with diffusely reflecting boundaries the sound attenuation along the length is greater than that with geometrically reflecting boundaries [2.9]. This corresponds to the qualitative analysis in Section 2.1.1 and the simulation results in Chapters 3 and 4.

Leschnik investigated the sound distribution in rectangular road tunnels using a computer model with the Monte-Carlo method [2.42]. The vehicles acted as sound sources as well as diffusers. It was found that the sound attenuation along the tunnel with a single source was greater when there were more vehicles in the tunnel. The computer model was validated by the measurements in a 1:20 scale model (5 m by 6.5 m by 40 m, full scale) and the site measurement in two tunnels (4·5 m by 6 m by 45 m and 5 m by 10 m by 1000 m). The importance of Leschnik's work was that the effectiveness of diffusers for the sound attenuation in long enclosures was experimentally demonstrated.

In urban streets, since there are always some irregularities on building or ground surfaces [2.43], a number of models have been proposed for taking diffuse reflections into account. In a model suggested by Davies, the sound field was assumed to be the sum of a multiple geometrically reflected field and a diffuse field that was fed from scattering at boundaries at each reflection of the geometrical field [2.44]. Wu and Kittinger [2.45], using Chien and Carroll's method of describing a surface by a mixed reflection law with an absorptivity or reflectivity parameter and a smoothness or roughness parameter [2.46], developed a model for predicting traffic noise. Heutschi suggested modelling sound propagation by a continuous energy exchange within a network of predefined points located on individual plane surfaces, and in this way it was possible to define any characteristic directivity pattern for the reflections [2.47]. Bullen and Fricke, in order to consider the effects of scattering from objects and protrusions in streets, analysed the sound field in terms of its propagating modes [2.48].

2.4.5.   Duct theory

Sound attenuation along ducts has been extensively investigated [2.49–2.52]. Although the dimension and boundary conditions are rather different, some of the theoretical principles for ducts are also useful for long spaces. The following semi-empirical formulae for calculating the sound attenuation along a duct are most commonly used [2.53–2.54]

$$D = K\bar{\alpha}\frac{U}{S} \tag{2.92}$$

$$D = 1 \cdot 5\alpha_n \frac{U}{S} \tag{2.93}$$

where D is the sound attenuation ratio in dB/m, K is a constant in relation to $\bar{\alpha}$, and α_n is the normal absorption coefficient.

It is noted that the above formulae are not directly applicable to long enclosures. By using equations (2.92) and (2.93) the attenuation along the length is linear, which is fundamentally different from the measurements in long enclosures [2.11,2.35,2.36 and also see Chapter 5]. Moreover, equations (2.92) and (2.93) are applicable at low frequencies ($\lambda \gg a, b$) and become inaccurate as the frequency increases [2.54]. Furthermore, equations (2.92) and (2.93) are valid for porous materials but are limited for structures with resonant absorption [2.55].

2.4.6.   Investigations on reverberation

The variation in reverberation time along the length was noticed by Clausen and Rekkebo when they carried out a series of measurements in a corridor [2.56]. Sergeev considered the reverberation in tunnels using the method of images [2.13]. The complexity of the result, however, made the work difficult to apply. By assuming that the source-receiver distance was small compared to the peripheral length of the cross-section, Hirata deduced an approximate theoretical formula for calculating the reverberation time in rectangular tunnels [2.57]. Barnett, by modifying Sabine's theory, gave an approximate formula for calculating the reverberation time in tube stations [2.58]. In the formula, however, the variation in reverberation time along the length was not taken into account.

Limited work has been carried out on reverberation in long spaces resulting from diffusely reflecting boundaries. As mentioned previously, Ollendorff considered reverberation in road tunnels using the partial differential equation of the diffusion of phonons [2.37]. Similarly, based on the analogy between a sound particle hitting walls and a single particle moving in a gas and hitting scattering objects, Picaut, Simon and Polack proposed a natural extension to the concept of diffuse sound field by introducing a new parameter, the diffusion coefficient [2.59]. One-dimensional solutions for rectangular enclosures with diffusely reflecting boundaries were given, and the increase in reverberation along the length was reported [2.60].

Reverberation in urban streets has also been paid attention. Kuttruff considered reverberation in two ways when he derived the formulae for the average value and the variance of noise level in urban streets [2.61]. For the first, the sound field was regarded as diffuse and the average absorption coefficient was calculated with a totally absorbent ceiling, hard walls and hard ground. For the second, the image source model was used and the side walls were considered to have an absorption coefficient. Steenackers, Myncke and Cops analysed the measured reverberation at a receiver that was close to the cross-section with source [2.62]. It was noticed that the decay curves were not straight. Reverberation in urban streets was also measured by several other researchers and the variation along the length was noticed [2.63,2.64].

In flat enclosures the unsuitability of classic room acoustic theory is similar to that in long enclosures [2.65–2.70]. It has been shown that in large factories, typically flat enclosures, the reverberation time is dependent on the distance from the source [2.71].

2.5.   A model for train noise prediction

In underground stations, typically long enclosures, train noise is usually a major acoustic problem — it causes unacceptable noise levels and reduces the speech intelligibility of PA systems. With train noise, instead of a single value, the intelligibility of a PA system becomes a dynamic process [2.72]. To predict the acoustic environment in underground stations and to investigate possible treatments for reducing train noise, it is necessary to establish a model to compute the temporal and spatial distribution of train noise. Such a model [2.73], TNS (train noise in stations), is described in this Section.

The logic of the modelling process is to consider a train as a series of sections and to calculate the train noise distribution in an underground station by inputting the sound attenuation along the length from a train section source in the underground system (i.e. the station and tunnel). The calculation sums the sound energy from all the train sections by considering train movement and the position of the train section in the train, and the input can be obtained by physical scale modelling.

2.5.1. Model train section and model train

To obtain the sound attenuation along the length from a train section by physical scale modelling, a simulation of the train section is essential. The physical scale model of a train section should be designed to simulate the shape, surface condition and the sound radiation characteristics of the actual train section. To investigate the effect of the train section when it is at different positions in a train, a physical scale model of the whole train is also necessary. Nevertheless, the simulation for the whole train, which is only for the shape and surface condition but not the sound radiation characteristics, is much simpler than that for the train section. In other words, in the scale model measurements the sound source is only the model train section but not the model train. In the following the 'physical scale model of a train section' and the 'physical scale model of a train' are called 'model train section' and 'model train' for short.

There is no need to specify the sound power level of the model train section. However, in order to obtain the absolute train noise level, the measurement data with the model train section should be calibrated by the difference in L_W between the model train and the actual train. This difference can be determined by measuring Lp_D, which is the difference in SPL between the model train and the actual train at the same relative receiver and in the same acoustic environment, preferably in a semi-free field (see Figure 2.24). The SPL of the actual train can be obtained by site measurement [2.74], and the SPL of the model train can be calculated using the measurement data of the model train section, as demonstrated below.

Consider a semi-free field. Assume that a model train of length of L_0 consists of N_0 train sections with equal distances between them. The SPL of the model train at receiver $(-L_0/2, y)$ can be determined by (see Figure 2.24(a))

$$Lp_M = 3 + 10\log \sum_{n=1}^{N_0/2} 10^{[Lp_n + S(n,v)]/10} \tag{2.94}$$

where Lp_n is the measured SPL when the model train section is at position n ($n = 1, \ldots, N_0/2$) of the model train. $S(n, v)$ is the difference between the L_W

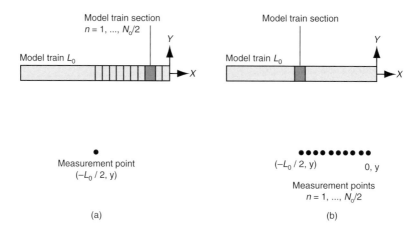

Figure 2.24. Measurement of the SPL of a model train in a semi-free field — plan view: (a) with $N_0/2$ positions of the model train section and one measurement point; (b) with one position of the model train section and $N_0/2$ measurement points

of the nth train section with a train velocity v and the reference train section with the reference train velocity v' (in TNS, $n = 1$ and $v' = 60\,\mathrm{km/h}$). In $S(n, v)$, the brake noise and door operation noise, which may have very high noise levels, should be considered. In addition, the noise radiated by rails, which is more noticeable than the airborne train noise when a train is far from the station, should be taken into account in $S(n, v)$.

The SPL of the model train can also be determined approximately using only one position of the model train section, as illustrated in Figure 2.24(b). In this case L_{p_n} in equation (2.94) is the measured SPL at measurement point n $(n = 1, \ldots, N_0/2)$.

A relationship between the train velocity v and the noise level S_T has been given [2.75], which is useful for determining $S(n, v)$

$$S_T = K_T + 10\log(v/v') \qquad\qquad (2.95)$$

where K_T is a constant in relation to room conditions and train types.

2.5.2. Database of a train section

With the above model train section and model train, the SPL distribution from a train section in an underground system (i.e. the station and tunnel) can be measured in a physical scale model of the system. Assume that the boundary conditions are the same along the station platform and also that they are the same along the tunnel. Then the sound attenuation between a train section and a receiver depends only on the distance between them along the length. By considering the train section being either in the station or in the tunnel,

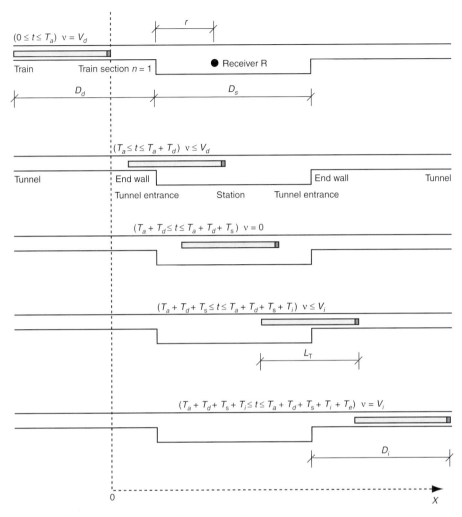

Figure 2.25. An underground system and the train movement — plan view

measurements should be carried out in the following three cases (see Figure 2.25):

(a) with the model train section and a microphone in the station, $Sss(d)$ (*ss*: station–station), where d is the distance between the model train section and the microphone;

(b) with the model train section in the tunnel and a microphone at the tunnel entrance, $Stt(d)$ (*tt*: tunnel–tunnel), where d is the distance between the model train section and the tunnel entrance — in other words, $Stt(d)$ is the SPL attenuation from the train section to the tunnel entrance; and

(c)   with the model train section in the tunnel and a microphone in the station, $Sts(d)$ (ts: tunnel–station), where d is the distance between the tunnel entrance and the microphone. In other words, $Sts(d)$ is the relative SPL attenuation from the tunnel entrance to the station.

With the three databases above, Sss, Stt and Sts, the spatial SPL distribution of a train section can be determined.

When the train section is in the tunnel, the effect of train noise can be obtained by considering both Stt and Sts, i.e. by summing the SPL attenuation from the train section to the tunnel entrance and the SPL attenuation from the tunnel entrance to the station. This is based on the assumption that the SPL at any position in the tunnel entrance is the same, which is true when the distance between the train section and the tunnel entrance is large enough. When establishing the database Sts by scale model measurements, this distance should be considered. An approximate database for Sts can also be measured using a loudspeaker with a size similar to that of the tunnel cross-section.

When the train section is in the station, the effect of train noise is obtained from Sss. To obtain a general database Sss, the model train section should be positioned far enough away from the hard end walls of the station to avoid reflections from the end walls. However, this may underestimate the SPL near the end walls when the train is in this range. More exact results in this range can be obtained using an extra Sss with the model train section near the end walls.

The distances between the receivers in the scale model are so chosen that the SPL at these points can represent the sound attenuation characteristics of the station and the tunnel with the required accuracy. Based on these data, in TNS more detailed databases can be produced automatically for finer distances.

In principle, the databases Sss, Stt and Sts could also be obtained by theoretical calculation. However, owing to the complexity of the source and a lack of effective theory, this would be inaccurate.

2.5.3.   Train movement

The above databases provide the spatial SPL distribution of a static train section. To obtain a temporal train noise distribution, it is necessary to consider the train movement. The movement of a train in an underground system is illustrated in Figure 2.25, in which it is assumed that the deceleration and acceleration are uniform. In this case, the train movement and the corresponding time range can be described as follows:

(a)   movement with velocity V_d, $0 \le t \le T_a$;
(b)   deceleration from velocity V_d to 0, $T_a \le t \le T_a + T_d$;
(c)   static at the middle of the station, $T_a + T_d \le t \le T_a + T_d + T_s$;
(d)   acceleration from velocity 0 to V_i, $T_a + T_d + T_s \le t \le T_a + T_d + T_s + T_i$; and

(e)   movement with velocity V_i, $T_a + T_d + T_s + T_i \leq t \leq T_a + T_d + T_s + T_i + T_e$.

Define the position of the first train section ($n = 1$) at time $t = 0$ as 0. Then, the position of this section at time t can be determined by

$$X_0 = V_d t \qquad (0 \leq t \leq T_a) \tag{2.96}$$

$$X_0 = V_d T_a + V_d(t - T_a) - \frac{V_d}{2T_d}(t - T_a)^2 \qquad (T_a \leq t \leq T_a + T_d) \tag{2.97}$$

$$X_0 = V_d T_a + \frac{V_d T_d}{2} \qquad (T_a + T_d \leq t \leq T_a + T_d + T_s) \tag{2.98}$$

$$X_0 = V_d T_a + \frac{V_d T_d}{2} + \frac{V_i}{2T_i}(t - T_a - T_d - T_s)^2$$
$$(T_a + T_d + T_s \leq t \leq T_a + T_d + T_s + T_i) \tag{2.99}$$

$$X_0 = V_d T_a + \frac{V_d T_d}{2} + \frac{V_i T_i}{2} + V_i(t - T_a - T_s - T_i - T_e)$$
$$(T_a + T_d + T_s + T_i \leq t \leq T_a + T_d + T_s + T_i + T_e) \tag{2.100}$$

Accordingly, at time t the position of the nth train section is

$$X_n = X_0 - (n-1)l_T \tag{2.101}$$

where l_T is the length of the train section. Correspondingly, the train velocity at time t, which is useful for determining $S(n, v)$, is

$$v = V_d \qquad (0 \leq t \leq T_a) \tag{2.102}$$

$$v = V_d - \frac{V_d}{T_d}(t - T_a) \qquad (T_a \leq t \leq T_a + T_d) \tag{2.103}$$

$$v = 0 \qquad (T_a + T_d \leq t \leq T_a + T_d + T_s) \tag{2.104}$$

$$v = \frac{V_i}{T_i}(t - T_a - T_d - T_s)$$
$$(T_a + T_d + T_s \leq t \leq T_a + T_d + T_s + T_i) \tag{2.105}$$

$$v = V_i \qquad (T_a + T_d + T_s + T_i \leq t \leq T_a + T_d + T_s + T_i + T_e) \tag{2.106}$$

2.5.4.   Noise from the whole train

Using the databases Sss, Stt and Sts, and considering the train movement above, the temporal and spatial noise distribution of a train section can be calculated. At time t the SPL caused by the nth train section at receiver R can be determined by

$$S_n = Stt(D_d - X_n) + Sts(r) + S(n, v) \qquad (0 \leq X_n < D_d) \tag{2.107}$$

$$S_n = Sss(|X_n - D_d - r|) + S(n, v) \qquad (D_d \leq X_n \leq D_d + D_s) \tag{2.108}$$

$$S_n = Sts(D_s - r) + Stt(X_n - D_d - D_s) + S(n, v)$$

$$(D_d + D_s < X_n \leq D_d + D_s + D_i) \quad (2.109)$$

where D_s is the length of the station, r is the distance between the receiver and the end wall (i.e. the tunnel entrance) where the train enters the station (see Figure 2.25), and D_d is the distance between the first train section and the end wall where the train enters the station when $t = 0$, and D_i is the distance between the last train section and the end wall where the train leaves the station when $t = T_a + T_d + T_s + T_i + T_e$. D_d and D_i can be calculated by

$$D_d = V_d T_a + \frac{V_d T_d}{2} - \frac{D_s + L_T}{2} \quad (2.110)$$

$$D_i = \frac{V_i T_i}{2} + V_i T_e - \frac{D_s + L_T}{2} \quad (2.111)$$

where L_T is the train length. Note that L_T is not necessarily the same as the full scale equivalent of L_0. The temporal and spatial noise distribution of the whole train can then be obtained by summing the sound energy from all the train sections. By considering the correction Lp_D, at time t the SPL caused by the whole train at receiver R is

$$Lp_T = 10 \log \sum_{n=1}^{N_T} 10^{S_n/10} + Lp_D \quad (2.112)$$

where N_T is the number of train sections in the train, $N_T = L_T/l_T$.

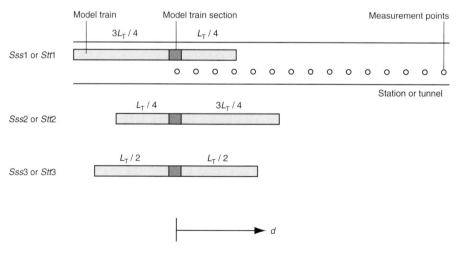

Figure 2.26. Scale model measurement of three groups of Sss or Stt by considering the effect of the train — plan view

2.5.5. Effect of train

The sound attenuation of a train section could be affected by its position in the train, especially when the cross-section of the station or tunnel is not considerably larger than that of the train. To consider this effect, measurements should be carried out by putting the model train section at various positions of the model train and, thus, a set of databases Sss, Stt and Sts can be established. The number of databases depends on the variation of these with different positions of the model train section. An example for establishing three Sss is illustrated in Figure 2.26, where the database that should be used is:

(a)   for $X_n < D_d + r$: $Sss1$ if $nl_T < L_T/4$, $Sss2$ if $L_T - nl_T < L_T/4$, and $Sss3$ otherwise;

(b)   for $X_n \geq D_d + r$: $Sss2$ if $nl_T < L_T/4$, $Sss1$ if $L_T - nl_T < L_T/4$, and $Sss3$ otherwise.

Stt can be established in a similar way.

An example for determining three Sts is shown in Figure 2.27. In this case, the database that should be used is:

(a)   for $X_n < D_d$: $Sts1$ if $X_0 < D_d$, $Sts3$ if $X_0 - D_d > L_T/2$, and $Sts2$ otherwise;

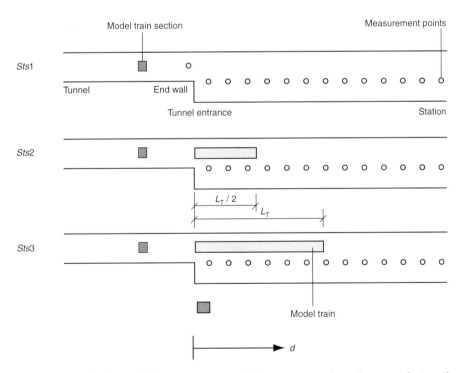

Figure 2.27. Scale model measurement of three groups of Sts by considering the effect of the train — plan view

(b)   for $X_n > D_d + D_s$: $Sts1$ if $X_0 - L_T > D_d + D_s$, $Sts3$ if $X_0 - L_T <$ $(D_d + D_s) - L_T/2$, and $Sts2$ otherwise.

In TNS, if a set of Stt, Sss and Sts is inputted, they can be selected automatically during the calculation. Examples of determining the databases in a scale model are given in Section 5.2.4.

2.5.6. TNS

The above calculations are included in the computer model TNS. The main inputs for the model are:

(a) Lp_D;
(b) $S(n, v)$;
(c) D_s, L_T, l_T;
(d) Stt, Sss, Sts; and
(e) T_a, T_d, T_s, T_i, T_e, V_d, V_i.

TNS can output the SPL of the train in the station with the following forms:

(a)   temporal distribution at various receivers;
(b)   spatial distribution at various times;
(c)   temporal distribution with frequency; and
(d)   temporal STI distribution at a given receiver — the calculation of the STI is described in Sections 1.5 and 2.6.

2.5.7. Validation

To validate TNS, a comparison between calculation and site measurement was made for London St John's Wood underground station. A detailed description of the station and its scale model is given in Section 5.2. The computation was carried out with the following inputs:

(a) $S(n, v) = S(1, 60) + 10 \log(v/60)$ (based on equation (2.95) and site measurement);
(b) $L_T = 75 \, \text{m}$, $N_T = 30 \, \text{m}$ and $D_s = 128 \, \text{m}$;
(c)   three Sss (cases 1, 3 and 6 in Figure 5.52);
(d)   three Sts (cases A, B and C in Figure 5.59);
(e) $Stt(d) = Sss(1 \cdot 5d)$ (theoretical estimation); and
(f)   uniform deceleration with $V_d = 72 \, \text{km/h}$ and $T_d = 21 \, \text{s}$ (based on site measurement).

Figure 2.28 shows the comparison of the temporal train noise distribution at 500 Hz, where $r = 7 \, \text{m}$ and the time range begins at the point at which the train

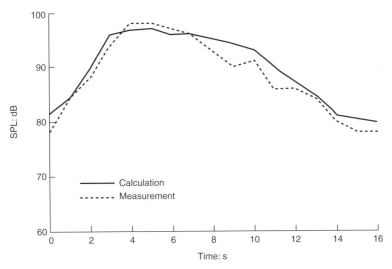

Figure 2.28. Comparison of the temporal train noise distribution at 500 Hz between calculation and site measurements in London St John's Wood underground station [2.7]

entered the station. It can be seen that the agreement between calculation and site measurement is very good — the accuracy is generally within ±2–3 dB.

2.6.   A model for predicting acoustic indices in long spaces

Due to the special acoustic properties in long spaces, an accurate and practical method for predicting the acoustic indices is required. Unfortunately, the practical use of the theoretical/simulation models presented in the above Sections is sometimes limited as relatively simple conditions are assumed due to the limitation of computing speed and the lack of appropriate theory. A better method for considering more complicated cases appears to be physical scale modelling but the use of this technique is often limited by the cost. Moreover, some factors like the loudspeaker directionality can be considered relatively easily in theoretical/simulation models but are difficult to simulate accurately in scale models. In this Section, ACL (acoustics of long spaces), a comprehensive model that combines the advantages of the both methods is presented [2.76].

The model first treats an actual long space as having a relatively simple geometry and calculates the acoustic indices of a single source using theoretical formulae or computer simulation models. The possible errors caused by the simplifications are then corrected using a database based on scale model measurements or site measurements. With the data of the single source, the acoustic indices of multiple sources can be calculated by summing the effect of each source. For speech intelligibility, the effect of ambient noise and the

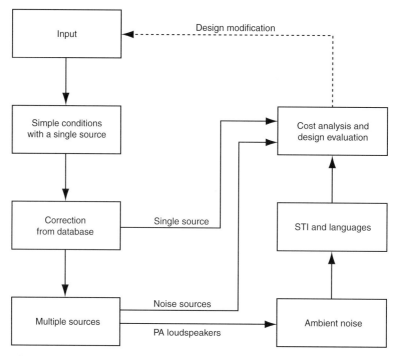

Figure 2.29. Block diagram of a method for predicting acoustic indices in long enclosures

differences among languages can be considered. Figure 2.29 shows the block diagram.

2.6.1. Input

The inputs of ACL are in six categories:

(*a*)   geometry, including dimensions and major obstructions;
(*b*)   boundary conditions, including absorption coefficient, absorber arrangements and diffuse situation;
(*c*)   source conditions, including source directionality, time delay and spatial distribution of sources — the sources can be noise sources or loudspeakers;
(*d*)   ambient noise for a PA system, where the input can take a form of temporal and spatial distribution;
(*e*)   proposed language(s) for a PA system; and
(*f*)   unit costs of the acoustic treatments.

2.6.2. Calculation of simple conditions

In ACL there are two ways of carrying out the calculation of simple conditions. One way is to use the theoretical formulae based on the method of images, as presented in Section 2.2. In this case the long spaces should be rectangular and with geometrically reflecting boundaries. Either a directional or an omnidirectional source can be considered. The other way is to use computer simulation models that are applicable to long spaces. In this case, the simplification of the actual long space can be made according to the model used.

The above calculation is for reverberation and SPL attenuation along the length — two fundamental acoustic indices in long spaces. It is essential to consider the reverberation variation along the length since this has been proved to be systematic. The reverberation at a receiver can be given by a decay curve, an energy response or, if the decay curve is approximately linear, by the EDT or RT30. The receiver spacing is so chosen that the reverberation and SPL at the receivers can represent the sound characteristics of the long space with a required accuracy. The calculation is for a single source. However, more than one source directionality should be considered if there are different source types.

2.6.3. Correction for more complicated cases

In some cases the above simplifications could bring considerable errors in calculation. For example, if there are diffusers on boundaries, the sound attenuation along a long space could be significantly underestimated by the assumption of geometrical reflection, as shown in Section 2.1.1 and Chapters 3 to 5. By using the diffuse model, however, the determination of the diffuse-reflection factor is often difficult [2.77], especially in a non-diffuse field, such as in a long space. Another factor that is difficult to consider accurately in theoretical/computer models is the effect of strategic arrangements of absorbers, especially when the absorbers are relatively small in size. With a given amount of absorption, strategic arrangements of absorbers could be effective for increasing the sound attenuation along the length and for decreasing reverberation, as shown in Chapter 5. As a result, it is necessary to correct the above calculation of simple conditions according to the actual situation.

A database for correction has been established in ACL based on intensive scale model tests. Currently the database is focused on the effects of diffuse reflection and strategic arrangements of absorbers. For the former, a ribbed diffuser and a Schroeder diffuser are considered (see Chapter 5). For the latter, 20 absorber arrangements are taken into account. As a principle of ACL, the database can always be built up from scale model or site measurements by considering more cases. These include more cross-section forms, more diffuser types and the effect of obstructions.

The correction is made by choosing an appropriate case from the database. Corresponding to the calculation of simple conditions, corrections are also made to the reverberation and SPL attenuation along the length. The correction to reverberation can be made for either decay curve or

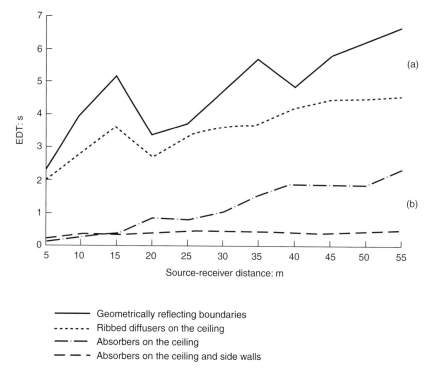

EDT: s

Source-receiver distance: m

———— Geometrically reflecting boundaries
------- Ribbed diffusers on the ceiling
—·— Absorbers on the ceiling
— — - Absorbers on the ceiling and side walls

Figure 2.30. Two examples of correction in EDT in a rectangular long enclosure: (a) effect of ribbed diffusers; (b) effect of absorber arrangements

reverberation time, depending on the calculation results above. It is noted that if more than one simplification is made, the simultaneous effects of these simplifications should be considered. For example, in long spaces the effectiveness of diffusers varies with absorption conditions, as demonstrated in Chapter 5. In the database, such simultaneous effects have been taken into account.

Two examples of correction are shown in Figures 2.30 and 2.31, where the data are the average of 500 Hz and 1000 Hz. The results are based on the measurements in a 1:16 scale model of a rectangular long space, as described in Section 5.1. The first example shows the effect of ribbed diffusers that are on the ceiling and with a density of 50%. It can be seen that in comparison with geometrically reflecting boundaries, with ribbed diffusers the EDT could be 30% shorter and the SPL attenuation along the length could be 3 dB greater. The second example is a comparison between two absorber arrangements. It is shown that for a given number of absorbers, when they are placed on both the ceiling and side walls, instead of on the ceiling only, the EDT could be 78% shorter and the SPL attenuation could be 5 dB greater.

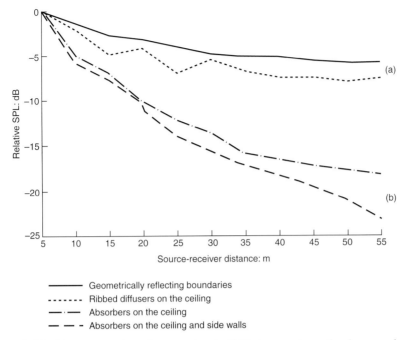

Figure 2.31. Two examples of correction in SPL in a rectangular long enclosure: (a) effect of ribbed diffusers; (b) effect of absorber arrangements

2.6.4. Multiple sources

Multiple sources are rather common in many long spaces, such as multiple loudspeakers of a PA system in underground stations, and vehicles in road tunnels and urban streets. With multiple sources the SPL distribution and reverberation in a long space become significantly different from those of a single source. In ACL, MUL (multiple sources in long spaces), a theoretical/ computer model for calculating acoustic indices with multiple sources is used as a sub-program [2.78,2.79].

Where there is a single source in a long space the energy response at a receiver can be described as $L(t)_z$, or $L(t)_{z,\psi}$ if the width is considerably larger than the height. Where z and ψ are the source-receiver distance along the length and width, respectively. $L(t)_z$ and $L(t)_{z,\psi}$ can be determined using the methods described in Sections 2.6.2 and 2.6.3. If there are different source types, data from each source type should be considered.

Consider $L(t)_z$. Assume that in a long space the end walls are totally absorbent, the boundary conditions along the length are all the same and all the sources have the same directionality. Then the sound propagation from a source to a receiver depends only on the distance along the length between them. In other words, the effect of any source on a given receiver can be determined from the data of a single source. With N_s sources the energy response at a receiver i can be calculated by

$$L_i(t) = K_W 10 \log \sum_{j=1}^{N_s} 10^{[L(t - D_j/c - t_j)_{z_j} + \Delta S_j]/10}$$

$$\left(t - \frac{D_j}{c} - t_j \geq 0 \right) \qquad (2.113)$$

where D_j is the distance between receiver i and the jth source, ΔS_j is the difference in L_W between the jth source and the reference source (normally $j = 1$) and t_j is the time delay of the jth source.

If the decay curves resulting from a single source are close to linear, equation (2.113) can, approximately, be simplified to

$$L_i(t) = K_W 10 \log \sum_{j=1}^{N_s} 10^{[L_{D_j} - (60/T_j)(t - D_j/c - t_j) + \Delta S_j]/10}$$

$$\left(t - \frac{D_j}{c} - t_j \geq 0 \right) \qquad (2.114)$$

where T_j and L_{D_j} are the RT and the steady-state SPL at receiver i resulting from the jth source. With equations (2.113) or (2.114) the reverberation time at receiver i from multiple sources can be determined.

The steady-state SPL at receiver i from all the sources can be calculated by

$$L_i = 10 \log \sum_{\Delta t} 10^{L_i(t)/10} - L_{ref} \qquad (2.115)$$

In a similar manner to $L(t)_z$, $L(t)_{z,\psi}$ can be considered.

In ACL the calculation from multiple sources is in two categories, namely multiple noise sources and multiple PA loudspeakers. For the former, the temporal and spatial distribution of multiple sources can be calculated and some corresponding indices, such as L_{eq}, can be given. For the latter, particular attention is paid to the speech intelligibility of multiple loudspeaker PA systems in underground stations.

2.6.5. Speech intelligibility in underground stations

In ACL the speech intelligibility of multiple loudspeaker systems in underground stations, measured by the STI and the RASTI, is calculated according to the method described in Section 1.6. For determining an optimal loudspeaker arrangement, calculation should be carried out with a range of loudspeaker spacings.

Ambient noise, especially train noise, is a vital factor when considering speech intelligibility in underground stations. In ACL, train noise is simulated using program TNS, as described in Section 2.5. In addition to train noise, ACL can also consider other noise sources, such as ventilation noise.

With a constant STI, the speech intelligibility could be different for different languages, as shown in Chapter 6. In ACL the word and sentence intelligibility of the proposed language(s) can be given corresponding to the calculated STI. Three languages have been included in ACL, namely English, Mandarin and Cantonese.

2.6.6.   Cost analysis and design evaluation

The final evaluation of an acoustic design for a long space depends on both acoustic indices and costs. A spreadsheet is given in ACL for estimating the cost of acoustic treatments, including initial costs and maintenance. If the design is unsatisfactory, modification can be made by adjusting the input.

2.6.7.   Validation

ACL has been validated by the measurements in Kwai Fong station, a typical Hong Kong MTR (mass transit railway) station.

2.6.7.1.   The station

At platform level the station was a typical rectangular long enclosure, as illustrated in Figure 2.32. Its length, width and height were 200 m, 20 m and 5.7 m, respectively. The up and down tracks were between two platforms. Major obstructions were four columns, two attendant cubicles and six escalators. It is noted that since the platform level was on the top floor, the obstructive effect of the escalators was not significant.

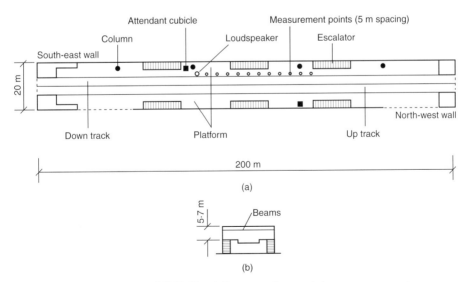

Figure 2.32. *Hong Kong MTR Kwai Fong station and the measurement arrangements: (a) plan; (b) cross-section*

The boundary conditions were approximately the same along the length. All the boundaries were concrete and there were no sound absorbent treatments on the boundaries. On either side of the north-west wall there was an open section of a length of 25 m. Both train entrances, each of height 6·2 m and width 8·8 m, were open to air. On the ceiling there were crossbeams with a spacing of about 2·4 m. Each beam was about 1·3 m high and 1·2 m wide. On the south-east wall there were continuous concrete panels of height 1·1 m and width 1·8 m. Each panel had a frame of 9 cm wide and 18 cm thick. Clearly, both the beams and frames should be regarded as diffusers, although the frames might only be effective at relatively high frequencies.

On each platform two lines of loudspeakers (Lamprodyn), one with an array of three loudspeakers at each position and the other with single loudspeakers, were arranged on the ceiling with a spacing of about 5 m. The major ambient noise was train noise. The PA messages were both in English and Cantonese.

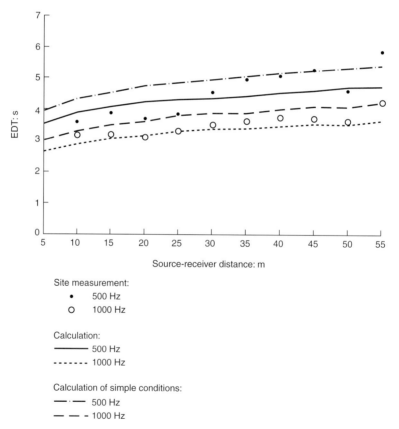

Figure 2.33. Comparison of EDT between site measurement and calculation in the Hong Kong MTR Kwai Fong station — the calculation of simple conditions, namely, without corrections, is also shown

2.6.7.2. Site measurement

A series of site measurements was carried out on the south-east platform. The single source measurement was made using a loudspeaker (Roland Type SB-20), which was positioned at a height of 1 m and a distance of 1·5 m from the platform edge. To avoid the effects of end walls and some obstructions, the loudspeaker was placed 66 m from a train entrance. The receivers were arranged along the platform, at a height of 1·25 m above the floor, a distance of 1·5 m from the platform edge and a spacing of 5 m. Measurements were carried out using MLSSA, a computer-aided system for architectural acoustic measurement [2.80]. Figures 2.33 and 2.34 show the measured EDT and SPL attenuation along the length at 500 Hz and 1000 Hz (octave). The RASTI from multiple PA loudspeakers was also measured, which was 0·34–0·38 along the south-east platform [2.81].

2.6.7.3. Calculation

The calculation of simple conditions is carried out using the theoretical formulae based on the method of images. The position and directionality of the source are in correspondence with the measurements described above. The calculation is made with $\alpha = 0·03$ at 500 Hz and $\alpha = 0·04$ at 1000 Hz, which appear to be a reasonable estimation of the boundary absorption [2.82]. The possible errors caused by ignoring the diffuse reflection of the

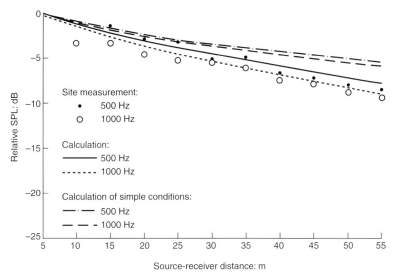

Figure 2.34. Comparison of SPL between site measurement and calculation in the Hong Kong MTR Kwai Fong station — the calculation of simple conditions, namely, without corrections, is also shown

ceiling and south-east wall are then corrected using the empirical database. The data used are from the measurements with ribbed diffusers in a scale model of an underground station that has similar geometry to the Kwai Fong station. The scale model is described in Section 5.1. The correction is based on the difference in the regression of measurement results with and without the model diffusers. In Figures 2.33 and 2.34 the calculated EDT and SPL attenuation at 500 Hz and 1000 Hz are shown, both with and without the correction.

Corresponding to the measurements, the calculated EDT increases with increasing source-receiver distance. As expected, the calculated EDT in simple conditions, namely, without corrections, is generally longer than the measured values, by about 10–25%.

With an approximate correction according to the measurement results with ribbed diffusers, the agreement between calculation and measurement is impressive and generally within ±10% accuracy with a few exceptions which are marginally outside. The correction is made by assuming that the diffuse reflection can bring 10–15% decrease in EDT, depending on source-receiver distance.

The predicted SPL attenuation along the length also shows very good agreement with the measured values, generally within ±1·5 dB accuracy. This is based on a correction with which it is assumed that the diffuse reflection can bring 0–3 dB extra SPL attenuation, depending on the source-receiver distance. Without corrections, as expected, the calculated SPL attenuation along the length is systematically less than the measured SPL attenuation, by about 2–4 dB at 30–55 m.

The calculation of the RASTI with multiple PA loudspeakers is carried out by assuming the boundary absorption coefficient as 0·02–0·08 at 250–8000 Hz. Corrections are made by considering the diffuse reflection of the ceiling and south-east wall. When the S/N ratio is 10–15 dB(A), which is approximately the same as that in the measurements, the calculated RASTI is 0·32–0·35 along the south-east platform, which is very close to the measured values. Corresponding to this RASTI range, the speech intelligibility is in the category of 'poor' for both English and Cantonese.

As expected, the calculation is sensitive to the absorption coefficient of the boundaries. By comparing Figures 2.33 and 2.34, it can be seen that the variation in EDT with different absorption coefficients is more significant than that in SPL attenuation. This is probably because the effect of a slight change in absorption coefficient can only become significant after many reflections, whereas the SPL mainly depends on early reflections.

In summary, from the above comparisons it is evident that ACL has predicted the acoustic indices in the station with a very good accuracy. Particularly, the usefulness of the correction from the empirical database has been demonstrated. This model would be useful for the acoustic design of new long spaces and for the investigation of possible acoustic treatments in existing long spaces.

2.7. References

2.1   KANG J. The unsuitability of the classic room acoustical theory in long enclosures. *Architectural Science Review*, 1996, **39**, 89–94.

2.2   KUTTRUFF H. *Room Acoustics*. Applied Science Publishers, Barking, England, 1991.

2.3   KANG J. Acoustic theory of long enclosures and its application to underground stations. *Underground Engineering and Tunnels*, 1998, No. 4, 37–40 (in Chinese).

2.4   KANG J. Acoustics in long underground spaces. *Tunnelling and Underground Space Technology*, 1997, **12**, 15–21.

2.5   KANG J. Acoustics in long enclosures. *Acustica/Acta Acustica*, 1996, **82**, S149.

2.6   KANG J. Reverberation in rectangular long enclosures with geometrically reflecting boundaries. *Acustica/Acta Acustica*, 1996, **82**, 509–516.

2.7   ORLOWSKI R. J. *London Underground Ltd Research — 99306 acoustic control systems for stations*. Arup Acoustics Report, No. AAc/46318/03, 1993.

2.8   KANG J. Sound attenuation in long enclosures. *Building and Environment*, 1996, **31**, 245–253.

2.9   KUTTRUFF H. Schallausbreitung in Langräumen. *Acustica*, 1989, **69**, 53–62.

2.10   KUNO K., KURATA T., NORO Y. and INOMOTO K. Propagation of noise along a corridor, theory and experiment. *Proceedings of the Acoustical Society of Japan*, 1989, 801–802 (in Japanese).

2.11   YAMAMOTO T. On the distribution of sound in a corridor. *Journal of the Acoustical Society of Japan*, 1961, **17**, 286–292 (in Japanese).

2.12   REDMORE T. L. A theoretical analysis and experimental study of the behaviour of sound in corridors. *Applied Acoustics*, 1982, **15**, 161–170.

2.13   SERGEEV M. V. Scattered sound and reverberation on city streets and in tunnels. *Soviet Physics — Acoustics*, 1979, **25**, 248–252.

2.14   SERGEEV M. V. Acoustical properties of rectangular rooms of various proportions. *Soviet Physics — Acoustics*, 1979, **25**, 335–338.

2.15   SAID A. Schalltechnische Untersuchungen im Strassentunnel. *Zeitschrift für Lärmbekämpfung*, 1981, **28**, 141–146.

2.16   SAID A. Zur Wirkung der schallabsorbierenden Verkleidung eines Strassentunnels auf die Spitzenpegel. *Zeitschrift für Lärmbekämpfung*, 1982, **28**, 74–78.

2.17   KANG, J. Sound propagation in street canyons: Comparison between diffusely and geometrically reflecting boundaries. *Journal of the Acoustical Society of America*, 2000, **107**, 1394–1404.

2.18   OLDHAM D. J. and RADWAN M. M. Sound propagation in city streets. *Building Acoustics*, 1994, **1**, 65–87.

2.19   IU K. K. and LI K. M. The propagation of sound in city streets. *Proceedings of the Seventh Western Pacific Regional Acoustics Conference*, Kumamoto, Japan, 2000, 811–814.

2.20   KANG J. Reverberation in rectangular long enclosures with diffusely reflecting boundaries. *Acustica/Acta Acustica* (to be published).

2.21   KANG J. Sound field resulting from diffusely reflecting boundaries: Comparison between various room shapes. *Proceedings of the 7th International Congress on Sound and Vibration*, Garmisch-Partenkirchen, Germany, 2000, 1687–1694.

2.22   STIBBS R. The prediction of surface luminances in architectural space. *Working Paper*, No. 54, Centre for Land Use and Built Form Studies, University of Cambridge, 1971.

2.23   FOLEY J. D., VAN DAM A., FEINER S. K. and HUGHES J. F. *Computer graphics: principle and practice*, 2nd edn. Addison-Wesley Publishing Company, 1990.

2.24   COHEN M. F. and GREENBERG D. P. The hemi-cube: A radiosity solution for complex environments. *Computer Graphics*, 1985, **19**, 31–40.

2.25   KANG J. A radiosity-based model for simulating sound propagation in urban streets. *Journal of the Acoustical Society of America*, 1999, **106**, 2262.

2.26   STEEMERS K., RAYDAN D. and KANG J. Evolutionary approach to environmentally conscious urban design. *Proceedings of the REBUILD*, Barcelona, Spain, 1999, 1–4.

2.27   KANG J. Sound field in urban streets with diffusely reflecting boundaries. *Proceedings of the Institute of Acoustics (IOA) (UK)*, Liverpool, 2000, **22**, No. 2, 163–170.

2.28   KANG J. Modelling the acoustic environment in city streets. *Proceedings of the PLEA 2000*, Cambridge, England, 512–517.

2.29   KANG J. Sound propagation in interconnected urban streets: a parametric study. *Environment and Planning B: Planning and Design*, 2001, **28**, 281–294.

2.30   YANG L. N. *Computer modelling of speech intelligibility in underground stations*. PhD Dissertation, University of South Bank, England, 1997.

2.31   YANG L. N. and SHIELD B. M. The prediction of speech intelligibility in underground stations of rectangular cross section. *Journal of the Acoustical Society of America*, 2001, **109**, 266–273.

2.32   THOMAS L. *Sound propagation in interconnected urban streets: a ray tracing model*. MSc dissertation, University of Sheffield, England, 2000.

2.33   RAHIM M. A. *Acoustic ray-tracing for a single urban street canyon with strategic design options*. MSc dissertation, University of Sheffield, England, 2001.

2.34   HOTHERSALL D. C., HOROSHENKOV K. V. and MERCY S. E. Numerical modelling of the sound field near a tall building with balconies near a road. *Journal of Sound and Vibration*, 1996, **198**, 507–515.

2.35   DAVIES H. G. Noise propagation in corridors. *Journal of the Acoustical Society of America*, 1973, **53**, 1253–1262.

2.36   REDMORE T. L. A method to predict the transmission of sound through corridors. *Applied Acoustics*, 1982, **15**, 133–146.

2.37   OLLENDORFF F. Diffusionstheorie des Schallfeldes im Strassentunnel. *Acustica*, 1976, **34**, 311–315.

2.38   MILES R. N. Sound field in a rectangular enclosure with diffusely reflecting boundaries. *Journal of Sound and Vibration*, 1984, **92**, 203–226.

2.39   GILBERT E. N. An iterative calculation of reverberation time. *Journal of the Acoustical Society of America*, 1981, **69**, 178–184.

2.40   KUTTRUFF H. A simple iteration scheme for the computation of decay constants in enclosures with diffusely reflecting boundaries. *Journal of the Acoustical Society of America*, 1995, **98**, 288–293.

2.41   KUTTRUFF H. Energetic sound propagation in rooms. *Acustica/Acta Acustica*, 1997, **83**, 622–628.

2.42   LESCHNIK W. Zur Schallausbreitung in einem Strassentunnel. *Deutsche Jahrestagung für Akustik*, DAGA 1976, Düsseldorf, 285–288.

2.43   LYON R. H. Role of multiple reflections and reverberation in urban noise propagation. *Journal of the Acoustical Society of America*, 1974, **55**, 493–503.

2.44   DAVIES H. G. Multiple-reflection diffuse-scattering model for noise propagation in streets. *Journal of the Acoustical Society of America*, 1978, **64**, 517–521.

2.45   Wu S. and KITTINGER E. On the relevance of sound scattering to the prediction of traffic noise in urban streets. *Acustica/Acta Acustica*, 1995, **81**, 36–42.

2.46   CHIEN C. F. and CARROLL M. M. Sound source above a rough absorbent plane. *Journal of the Acoustical Society of America*, 1980, **67**, 827–829.

2.47   HEUTSCHI K. Computermodell zur Berechnung von Bebauungszuschlägen bei Straßenverkehrslärm. *Acustica/Acta Acustica*, 1995, **81**, 26–35.

2.48   BULLEN R. and FRICKE F. Sound propagation in a street. *Journal of Sound and Vibration*, 1976, **46**, 33–42.

2.49   RAMSEY J. R. *Architectural, building and mechanical systems acoustics — A guide to technical literature*. Lacrosse, USA, 1986.

2.50   MECHEL F. Einfluss der Querunterteilung von Absorbern auf die Schallausbreitung in Kanälen. *Acustica*, 1965/66, **16**, 90–100.

2.51   DOAK P. E. Fundamentals of aerodynamic sound theory and flow duct acoustics. *Journal of Sound and Vibration*, 1973, **28**, 527–561.

2.52   BAXTER S. M. and MORFEY C. L. Modal power distribution in ducts at high frequencies. *AIAA Journal*, 1983, **21**, 74–80.

2.53   SABINE J. H. The absorption of noise in ventilating ducts. *Journal of the Acoustical Society of America*, 1940, **12**, 53–57.

2.54   MAA D. Y. and SHEN H. *Handbook of Acoustics*. Science Press, Beijing, 1987 (in Chinese).

2.55   PIAZZA R. S. The attenuation in ducts lined with selective structures. *Acustica*, 1965, **15**, 402–407.

2.56   CLAUSEN J.-E. and REKKEBO J. A. *Acoustical properties of corridors*. Siv. ing. Thesis. The Norwegian Institute of Technology, 1976 (in Norwegian).

2.57   HIRATA Y. Geometrical acoustics for rectangular rooms. *Acustica*, 1979, **43**, 247–252.

2.58   BARNETT P. W. Acoustics of underground platforms. *Proceedings of the Institute of Acoustics (UK)*, 1994, **16**, No. 2, 433–443.

2.59   PICAUT J., SIMON L. and POLACK J.-D. A mathematical model of diffuse sound field based on a diffusion equation. *Acustica/Acta Acustica*, 1997, **83**, 614–621.

2.60   PICAUT J., SIMON L. and POLACK J.-D. Sound field in long enclosures with diffusely reflecting boundaries. *Proceedings of the 16th International Congress on Acoustics (ICA)*, Seattle, USA, 1998, 2767–2768.

2.61   KUTTRUFF H. Zur Berechnung von Pegelmittelwerten und Schwankungsgrößen bei Straßenlärm. *Acustica*, 1975, **32**, 57–69.

2.62   STEENACKERS P., MYNCKE H. and COPS A. Reverberation in town streets. *Acustica*, 1978, **40**, 115–119.

2.63   YEOW K. W. External reverberation times observed in built-up areas. *Journal of Sound and Vibration*, 1976, **48**, 438–440.

2.64   KO N. W. M. and TANG C. P. Reverberation time in a high-rise city. *Journal of Sound and Vibration*, 1978, **56**, 459–461.

2.65   HODGSON M. R. *Theoretical and physical models as tools for the study of factory sound fields*. PhD dissertation, University of Southampton, 1983.

2.66   ORLOWSKI R. J. The arrangement of sound absorbers for noise reduction: results of model experiments at 1:16 scale. *Noise Control Engineering Journal*, 1984, **22**, 54–60.

2.67   HODGSON M. R. and ORLOWSKI R. J. Acoustic scale modelling of factories: Part I — Background, instrumentation and procedures. *Journal of Sound and Vibration*, 1987, **113**, 29–46.

2.68   HODGSON M. R. On the prediction of sound fields in large empty rooms. *Journal of the Acoustical Society of America*, 1998, **84**, 253–261.

2.69   DANCE S. M. and SHIELD B. M. Factory noise computer prediction using a complete image-source method. *Proceedings of Euro-noise '92*, London, 1992, 458–492.

2.70   AKIL H. A. and OLDHAM D. J. Determination of the scattering parameters of fittings in industrial buildings for use in computer based factory noise prediction models: Part I — Theoretical background. *Building Acoustics*, 1995, **2**, 461–481.

2.71   VIGRAN T. E. Reverberation time in large industrial halls. *Journal of Sound and Vibration*, 1978, **56**, 151–153.

2.72   KANG J. A method for predicting train noise in subway stations. *Journal of the Acoustical Society of America*, 1996, **100**, 2587.

2.73   KANG J. Modelling of train noise in underground stations. *Journal of Sound and Vibration*, 1996, **195**, 241–255.

2.74   SHIELD B. M., ROBERTS J. P. and VUILLERMOZ M. L. Noise and the Docklands light railway. *Applied Acoustics*, 1989, **26**, 305–315.

2.75   HARRIS C. M. *Handbook of Noise Control*. McGraw-Hill, New York, 1979.

2.76   KANG J. A method for predicting acoustic indices in long enclosures. *Applied Acoustics*, 1997, **51**, 169–180.

2.77   LAM Y. M. The dependence of diffusion parameters in a room acoustics prediction model on auditorium sizes and shapes. *Journal of the Acoustical Society of America*, 1996, **100**, 2204–2212.

2.78   KANG J. Acoustics in long enclosures with multiple sources. *Journal of the Acoustical Society of America*, 1996, **99**, 985–989.

2.79   KANG J. Speech intelligibility of multiple loudspeaker public address systems in long enclosures. *Journal of the Acoustical Society of America*, 1995, **98**, 2982.

2.80   DRA LABORATORIES. *MLSSA. Reference Manual Version 9.0*. Sarasota, Florida, 1994.

2.81   HO C. L. *Station public address — acoustic measurement study*. Hong Kong Mass Transit Railway Corporation Report, ODE/D-TEL/C4246/01, 1992.

2.82   INSTITUTE OF BUILDING PHYSICS. *Handbook of Architectural Acoustics. China Building Industry Press, Beijing*, 1985 (in Chinese).

3. Parametric study: design guidance for long enclosures

Based on the theoretical/computer models described in Chapter 2, a series of parametric studies for long enclosures is presented in this Chapter, aiming at providing guidance for practical design. Major parameters considered include cross-sectional area and aspect ratio, amount and distribution of absorption, reflection characteristics of boundaries, number and position of sources, source directionality and air absorption. The sound attenuation along the length is analysed in Section 3.1, and the characteristics of reverberation are examined in Sections 3.2 and 3.3. In Section 3.4 the sound behaviour of multiple sources is studied.

3.1. Sound attenuation along the length

The calculation is carried out by assuming point source and rectangular cross-section [3.1]. The image source method (Section 2.2) and Davies' method (Section 2.4.2) are used for geometrically reflecting boundaries and the radiosity method (Section 2.3) is applied to diffusely reflecting boundaries. In the calculation, the range of cross-sectional area and aspect ratio is representative of actual tunnels, in particular, some underground stations. Usually corridors also fall within this range. In principle, these dimensions are within the applicable range of the above three methods, although the calculation results with highly absorbent boundaries at long distances may only be interpreted relatively to one another rather than absolutely, as analysed in Section 2.2.2. For the sake of convenience, air absorption is not considered in this Chapter except where indicated.

There are two kinds of sound attenuation. One is the relative attenuation with reference to a given distance from the source and the other is the absolute attenuation with reference to the source power level, L_W. In Davies' formulae, the former is in correspondence with the attenuation of the total energy in the whole cross-section, and the latter with the attenuation of the energy per unit area in a cross-section.

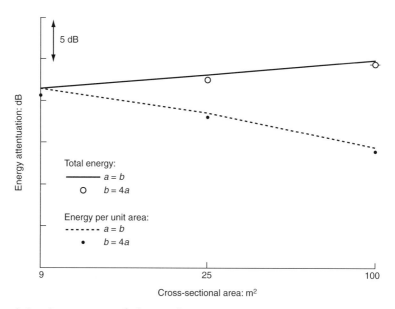

Figure 3.1. Attenuation of the total energy in the whole cross-section and the energy per unit area in a cross-section — α = 0·05 and the receiver is at 120 m, calculated using equation (2.82)

3.1.1.   Cross-sectional area and aspect ratio

The attenuation of the total energy in the whole cross-section, namely $10 \log (P_{SO}/P_{IN})$ obtained using equation (2.82) (the simple source), is shown in Figure 3.1. Six long enclosures with three cross-sectional areas ($9 \, m^2$, $25 \, m^2$ and $100 \, m^2$) and two aspect ratios (1 : 1 and 4 : 1) are considered, namely 3 m by 3 m, 6 m by 1·5 m, 5 m by 5 m, 10 m by 2·5 m, 10 m by 10 m and 20 m by 5 m. The receiver is at 120 m and the boundary absorption coefficient is 0·05. It can be seen that the larger the cross-section is, the less the attenuation is. This is mainly because the number of reflections in a given time is less for a larger cross-section.

Contrary to the attenuation of the total energy in the whole cross-section, Figure 3.1 also shows that the attenuation of the energy per unit area in a cross-section, namely $10 \log (9P_{SO}/SP_{IN})$, is greater with a larger cross-section, where the unit area is assumed as 3 m by 3 m. This result appears closer to the subjective loudness.

The calculation based on equation (2.26), namely the image source method, is shown in Figure 3.2. The configurations correspond to those in Figure 3.1. The source and receivers are along the centre of the cross-section. It can be seen that with increasing cross-sectional area the relative SPL attenuation from 20 m to 120 m becomes less, but the SPL attenuation at 120 m with reference to L_W becomes greater. In other words, with a larger cross-sectional

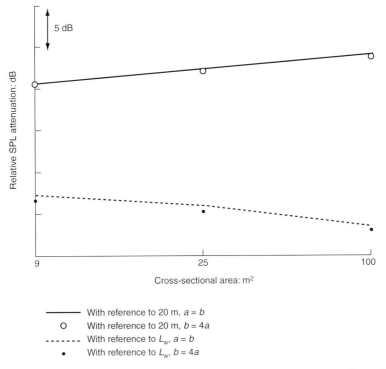

Figure 3.2. Relative SPL attenuation at 120 m with geometrically reflecting boundaries — α = 0·05, calculated using equation (2.26)

area, the relative attenuation with reference to a given distance from the source is less but the absolute attenuation with reference to the source power is greater. This is in correspondence with the results shown in Figure 3.1.

For the same configurations, the calculation using the radiosity model is shown in Figure 3.3. In terms of the relative attenuation with reference to a given distance from the source, the result with diffusely reflecting boundaries is similar to that with geometrically reflecting boundaries. In terms of the absolute attenuation with reference to the source power, however, the results with the two types of boundary are opposite. In Figure 3.3 it is seen that with a larger cross-section the SPL attenuation at 120 m with reference to L_W becomes less. This is probably because the overall reflection order at the receiver becomes less with a larger cross-section. Another possible explanation is that, with increasing cross-sectional area the sound field becomes closer to diffuse and, thus, the attenuation is less.

The effect of the cross-sectional aspect ratio can be seen in Figures 3.1 to 3.3. With a constant cross-sectional area, when the width/height ratio is $4:1$, namely $b = 4a$, the sound attenuation tends to be slightly greater than that of the square section.

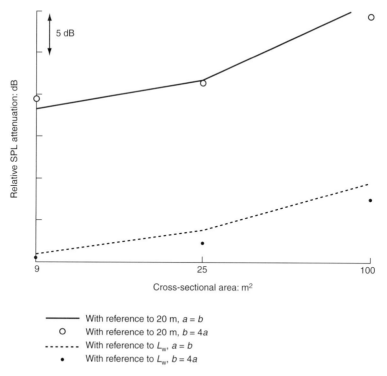

Figure 3.3. Relative SPL attenuation at 120 m with diffusely reflecting bound-aries — α = 0·05, calculated using the radiosity model

3.1.2. Amount of boundary absorption

The effect of boundary absorption on the SPL attenuation is shown in Figure 3.4, where $S = 3$ m by 3 m, the source and receiver are along the centre of the cross-section, the source-receiver distance is 50 m and the SPL is with reference to L_W. Both geometrically and diffusely reflecting boundaries are considered and the calculations are based on the image source method and the radiosity method, respectively. It can be seen that for both kinds of boundary, as the absorption coefficient increases linearly from 0·05 to 0·5, the attenuation increases with a decreasing gradient. In other words, the efficiency of absorbers per unit area is greater when there are fewer absorbers. This phenomenon, which has also been noticed in regularly-shaped (i.e. quasi-cubic) enclosures [3.2], appears physically reasonable.

Using equations (2.83) and (2.85) (the equal energy source), calculation is carried out by increasing the number of absorbent boundaries. Figure 3.5 shows the variation of the SPL attenuation at 50 m and 120 m with reference to P_{IN} when absorbers of $α = 0·8$ are arranged on one to four boundaries. In the calculation the absorption coefficient of the other boundaries is 0·05 and $S = 3$ m by 3 m. Figure 3.5 shows that as the number of the absorbent boundaries increases, the attenuation increases with a decreasing gradient. In principle, this is similar to the results shown in Figure 3.4.

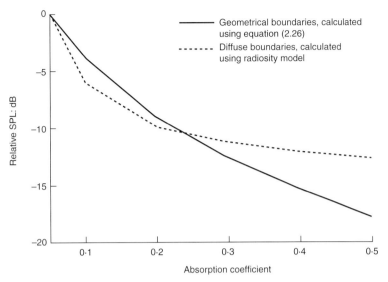

Figure 3.4. Sound attenuation at 50 m with reference to L_W as the absorption coefficient increases

3.1.3. Distribution of boundary absorption
The SPL attenuation in a long enclosure with various distributions of absorption in cross-section is shown in Figure 3.6, where $S = 5$ m by 5 m, the source and receiver are along the centre of the cross-section, the source-receiver

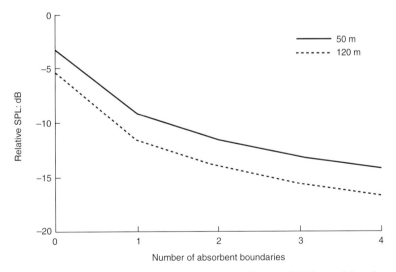

Figure 3.5. Variation of the SPL attenuation at 50 m and 120 m with reference to P_{IN} as the number of absorbent boundaries ($\alpha = 0.8$) increases from one to four — $S = 3$ m by 3 m, calculated using equations (2.83) and (2.85)

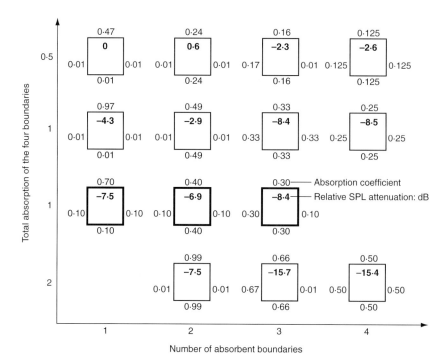

Figure 3.6. Normalised relative SPL attenuation at 120 m with reference to L_W with various distributions of absorption in cross-section — the boundaries are geometrically reflective and $S = 5$ m by 5 m, calculated using equation (2.29)

distance is 120 m, and the SPL is with reference to L_W and is normalised by the SPL in the distribution with $\alpha_F = 0.01$, $\alpha_C = 0.47$ and $\alpha_{A,B} = 0.01$. The calculation is carried out using equation (2.29), a formula based on the image source method. It can be seen that the attenuation is significantly higher when the absorbers are on three or four boundaries. A possible reason for this is that with more than one hard boundary, especially when two of them are parallel, it is possible for some reflections to reach the receiver without impinging upon any absorbent boundaries. For this reason, in the case of one or two absorbent boundaries, the attenuation is significantly greater when the absorption coefficient of the hard walls becomes 0·1 from 0·01, as shown in Figure 3.6. With a given amount of total absorption, the attenuation with three or four absorbent boundaries is nearly the same.

The situation with diffusely reflecting boundaries is rather different. With the same configurations as above, calculation using the radiosity method is shown in Figure 3.7, where two source-receiver distances, 20 m and 120 m, are considered. On the one hand, it can be seen that in comparison with Figure 3.6, the differences in sound attenuation between various absorption distributions become much less. This is probably because with diffusely reflecting boundaries

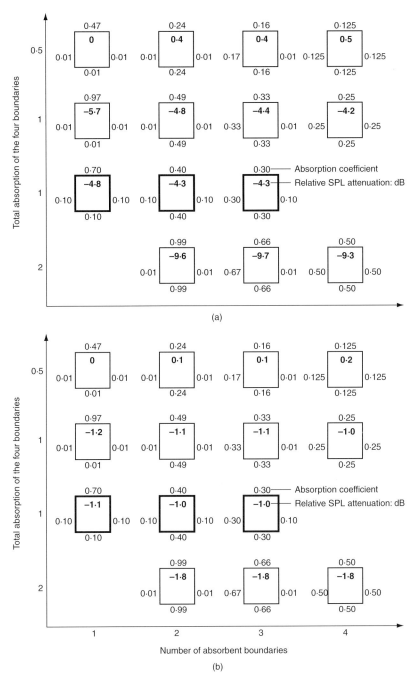

Figure 3.7. Normalised relative SPL attenuation with reference to L_W with various distributions of absorption in cross-section. The boundaries are diffusely reflective and S = 5 m by 5 m. Calculated using the radiosity method: (a) source-receiver distance 20 m; (b) source-receiver distance 120 m

a sound ray has more chances to hit all the boundaries. On the other hand, Figure 3.7 shows a slight tendency that the sound attenuation along the length is the highest if certain absorbers are arranged on one boundary and the lowest if they are evenly distributed on all boundaries. A simplified explanation is given in Section 3.3.4.

3.1.4. Comparison between diffusely and geometrically reflecting boundaries

From the above Sections it can be seen that there are noticeable differences between sound fields in long enclosures resulting from geometrically and diffusely reflecting boundaries. Figure 3.8 shows a comparison in relative sound attenuation along the length between the two kinds of boundary. In the calculation $L = 60$ m, $S = 6$ m by 4 m, $\alpha = 0.2$, and the two end walls are totally absorbent. A point source is at one end of the long enclosure and at the centre of the cross-section. The receivers are also along the centre of the long enclosure, and the source-receiver distance is 1–60 m. From Figure 3.8 it is important to note that with geometrically reflecting boundaries the SPL attenuation is considerably less than that with diffusely reflecting boundaries. This is in correspondence with the theoretical results by Kuttruff, as mentioned in Section 2.4.4, and with experimental results, as described in Chapter 5. A more detailed comparison between the two kinds of boundary is given in Section 4.2.

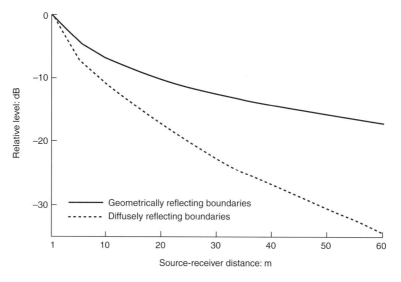

Figure 3.8. Comparison of the relative sound attenuation along the length between geometrically and diffusely reflecting boundaries

3.2. Reverberation resulting from geometrically reflecting boundaries

First, this Section analyses basic characteristics of reverberation in long enclosures, especially the variation along the length. It then discusses the effects of air absorption, cross-sectional area and aspect ratio, source directionality and end walls [3.3]. The calculation is based on the image source method described in Section 2.2.

Particular attention is paid to the EDT. This is because the early decay of a long enclosure governs the STI. Measurements in five London underground stations show that the correlation coefficient between the STI and the full band EDT (125–4000 Hz) is 0·92 (linear regression) or 0·94 (logarithmic regression), whereas the correlation coefficient between the STI and the full band RT30 is only 0·77 (linear regression) or 0·78 (logarithmic regression) [3.4].

3.2.1. Variation in reverberation along the length

Figure 3.9 shows the variation of RT30 and the EDT along the length of an infinitely long enclosure of $S = 6$ m by 4 m. In the calculation, two absorption coefficients, $\alpha = 0\cdot1$ and $0\cdot2$, are considered. It is interesting to note that with

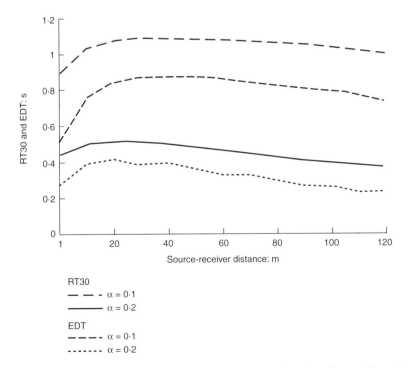

Figure 3.9. Variation of RT30 and EDT along the length of an infinitely long enclosure — $S = 6$ m by 4 m

the increase of source-receiver distance, the RT30 and EDT increase rapidly until a maximum and then decrease slowly. This is caused mainly by two opposing factors, namely the relative change in path length of reflected sound and the average order of image sources (i.e. the number of reflections) at a given time after the direct sound.

In one case, with increasing source-receiver distance the relative change in path length from time t to $t + \Delta t$ becomes less (see Figure 2.1(b)), so that the rate of decay is less and the reverberation time is longer. In the other case, using equations (2.9) and (2.10), it is evident that at a given time after the direct sound the average order of image sources is higher for a greater source-receiver distance, and this becomes more significant with increasing time. Given the fact that with a higher order of image sources more energy is absorbed by the boundaries, with increasing source-receiver distance the rate of decay becomes greater and the reverberation time becomes shorter.

Within a certain source-receiver distance, the first factor is dominant and, thus, the reverberation time increases along the length. This is particularly significant if the early part of the decay curve is considered, which can be seen by comparing EDT and RT30 in Figure 3.9. The reason is that the relative change in path length becomes less with increasing time. Beyond a certain point, conversely, the second factor becomes more dominant and, thus, the reverberation time decreases along the length. This is particularly significant for a higher α since, in this case, the order of image sources is more effective. The effect of α is seen in Figure 3.9 by comparing $\alpha = 0.1$ and 0.2. Correspondingly, with a higher α the point where the reverberation time decreases is closer to the source.

Besides the above two factors, the number of image sources also affects the variation in reverberation time along the length. From equations (2.8) to (2.11) it is evident that this effect is similar to that of the image source order, but its effectiveness is much less.

The RT30/EDT ratios corresponding to Figure 3.9 are shown in Figure 3.10. It is seen that these ratios are all greater than 1, especially in the near field. In other words, the decay curves are not linear. This is mainly because in long enclosures the increase in the number of reflections with time is slower than that in a diffuse field.

3.2.2. Air absorption

The reduction of sound energy caused by air absorption increases with time, since the effect of air absorption is greater with a longer reflection distance. This means that with air absorption the decay curves are more linear. Figure 3.11 shows a comparison of decay curves with $M = 0.1\,\mathrm{dB/m}$ and $M = 0$ (without air absorption) in an infinitely long enclosure where $S = 6\,\mathrm{m}$ by $4\,\mathrm{m}$, $z = 50\,\mathrm{m}$ and $\alpha = 0.1$. When $M = 0.1$, the RT30/EDT ratio is 1.2, which is

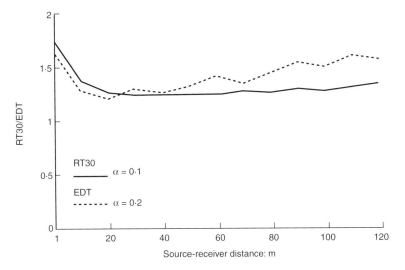

Figure 3.10. RT30/EDT ratios corresponding to Figure 3.9

less than that of $M = 0$, when the RT30/EDT ratio is 1·3. Correspondingly, for $z = 3$ m, the RT30/EDT ratio is 1·5 for $M = 0·1$ and 1·6 for $M = 0$.

3.2.3. Cross-sectional area and aspect ratio

Figure 3.12 compares the EDT in five infinitely long enclosures with different cross-sections: 12 m by 8 m, 16 m by 6 m, 24 m by 4 m, 6 m by 4 m and 8 m by 8 m, where $\alpha = 0·1$.

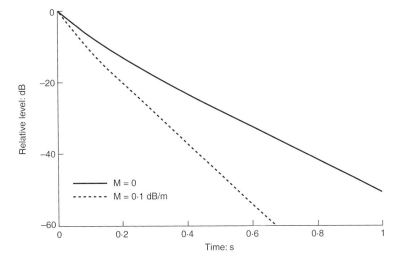

Figure 3.11. Decay curves in an infinitely long enclosure with $M = 0$ and $M = 0·1 \, dB/m$ — $S = 6 \, m$ by $4 \, m$, $z = 50 \, m$ and $\alpha = 0·1$

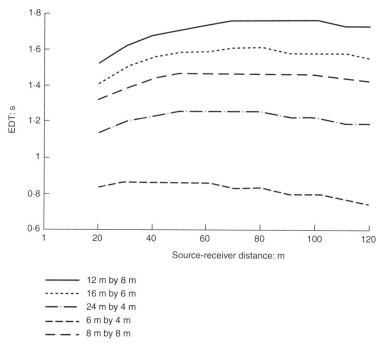

EDT: s

Source-receiver distance: m

——— 12 m by 8 m
······· 16 m by 6 m
—·— 24 m by 4 m
– – – 6 m by 4 m
— — · 8 m by 8 m

Figure 3.12. EDT in five infinitely long enclosures with different cross-sections: 12 m by 8 m, 16 m by 6 m, 24 m by 4 m, 6 m by 4 m and 8 m by 8 m — α = 0·1

With a constant aspect ratio, the reverberation time is longer for a larger cross-section (compare $S = 6$ m by 4 m and 12 m by 8 m in Figure 3.12). This is similar to the situation in the diffuse sound field given that a larger cross-section corresponds to a larger room volume. However, unlike the results obtained by classic formulae, the increase of reverberation time with increasing cross-sectional area is more significant for a longer source-receiver distance. Moreover, calculations show that for a larger cross-section the RT30/EDT ratio is greater, especially in the near field. This is because the difference in the sound path length between various image sources is greater for a larger cross-section.

For a given cross-sectional area, using equations (2.9) and (2.11) it is evident that the reverberation time reaches a maximum as the aspect ratio tends to 1. The main reason is that the average order of image sources decreases as the cross-section tends towards square. This phenomenon can be seen in Figure 3.12 by comparing $S = 12$ m by 8 m, 16 m by 6 m and 24 m by 4 m. Due to the effect of aspect ratio, the reverberation time may be longer with a smaller cross-sectional area but a square cross-section. Such an example is shown in Figure 3.12 (compare $S = 24$ m by 4 m and 8 m by 8 m).

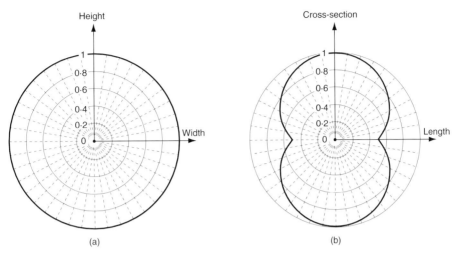

Figure 3.13. Intensity radiation of an imaginary directional source: (a) cross-section; (b) length-wise section

3.2.4. Directional source

To investigate the effect of source directionality, a directional source is assumed (see Section 2.2.1.2), as shown in Figure 3.13, where

$$f(\beta) = \tfrac{1}{2}\left[\sin(\beta) + 1\right] \tag{3.1}$$

Figure 3.14 shows a comparison of the RT30 and EDT between the directional source and a point source in an infinitely long enclosure where

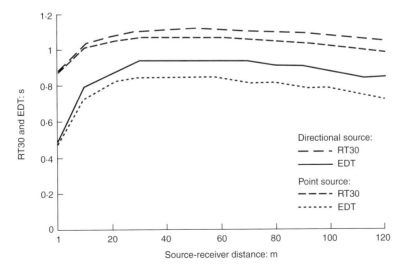

Figure 3.14. Comparison of reverberation time between the directional source and a point source in an infinitely long enclosure — S = 6 m by 4 m and α = 0·1

$S = 6\,\text{m}$ by $4\,\text{m}$ and $\alpha = 0\cdot1$. The calculation is carried out using equation
(2.13). It can be seen that with this directional source the RT30 and EDT
are systematically longer. This is due to the increase of initial sound energy
for later reflections.

This example demonstrates the possibility of varying reverberation time in
long enclosures using a directional source. A similar example is shown in Section
3.4.2. Correspondingly, the sound attenuation along the length can also be
changed with a directional source. For instance, in the above case the SPL of
the directional source at $z = 80\,\text{m}$ is 1.3 dB lower than that of the point source.

3.2.5. End wall

Figure 3.15 shows the $L(t)$ curves with $\alpha_e = 0\cdot1$, $0\cdot8$ and 1 in a finite long
enclosure, where $S = 6\,\text{m}$ by $4\,\text{m}$, $L = 120\,\text{m}$, $z = 20\,\text{m}$, $\alpha = 0\cdot1$, and the distance between the source and an end wall is 60 m. The calculation is made
using equation (2.16) by considering eight image source planes. It can be
seen that the end walls can have a significant effect on the $L(t)$ curves. When
α_e is $0\cdot1$, the end walls can bring more than 20 dB increase on the curves.
This means that highly absorbent end walls can be very helpful for reducing
the reverberation time in long enclosures.

The effect of end walls on the sound attenuation along the length is not as
significant as on the reverberation. For example, in Figure 3.15, in comparison
with the infinite case, the sound attenuation is only 1 dB less when $\alpha_e = 0\cdot1$.
This is because the SPL depends mainly on early reflections, whereas end
walls are normally effective for multiple reflections.

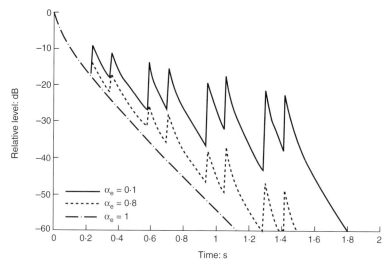

Figure 3.15. L(t) curves in a finite long enclosure with $\alpha_e = 0\cdot1$, $0\cdot8$ and 1 —
S = 6 m by 4 m, L = 120 m, d_s = 60 m, z = 20 m and $\alpha = 0\cdot1$

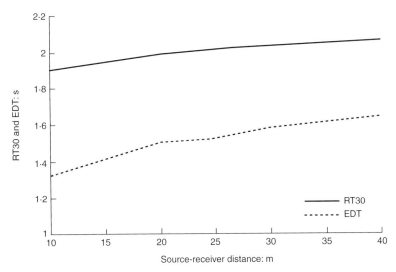

Figure 3.16. RT30 and EDT in an auditorium-like enclosure — S = 10 m by 25 m, L = 40 m, α = 0·15 and α_e = 1

3.2.6. An auditorium-like enclosure

Figure 3.16 shows the calculation for an enclosure with $S = 10$ m by 25 m, $L = 40$ m and $\alpha_e = 1$. This is an approximate simulation of some auditoriums where the stage and end wall are strongly absorbent. For the sake of convenience, it is assumed that the ceiling, floor and side walls all have an absorption coefficient of 0·15.

In this auditorium-like enclosure the RT30 and EDT increase along the length. This may explain why in the front part of some auditoriums the sound is relatively 'dry'. The reverberation time calculated using the Eyring formula is 1·49 s, which is fundamentally different from the results in Figure 3.16.

3.3. Reverberation resulting from diffusely reflecting boundaries

The calculation in this Section is based on the radiosity method described in Section 2.3. In addition to analysing basic characteristics of reverberation resulting from diffusely reflecting boundaries, comparison is made between geometrically and diffusely reflecting boundaries. Similar to the previous Section, effects of air absorption, cross-sectional area and aspect ratio, distribution of absorption and end walls are examined [3.5].

3.3.1. Reverberation along the length

To examine the variation in reverberation along long enclosures with diffusely reflecting boundaries, calculation is made with a typical configuration, namely

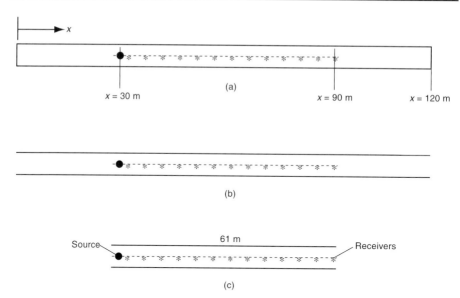

Figure 3.17. Plan view of three long enclosures showing the source position and receiver arrangements: (a) case 1; (b) case 2; (c) case 3

$L = 120$ m and $S = 6$ m by 4 m. The calculation arrangement is illustrated in case 1 of Figure 3.17. For the sake of convenience, it is assumed that the boundary absorption is uniform and the absorption coefficient is 0·2. An omnidirectional source is at (30 m, 3 m, 2 m) (see Figure 2.16). The receivers are along the centre of the long enclosure and the source-receiver distance is 1–60 m.

The calculated RT30 and EDT along the length are shown in Figure 3.18. It can be seen that the reverberation varies considerably along the length. In the calculated range, with the increase of source-receiver distance, the RT30 increases continuously and the EDT increases rapidly until it reaches a maximum and then decreases slowly. In Figure 3.19 the decay curves at four typical source-receiver distances, namely 1 m, 5 m, 20 m and 55 m, are given. From Figures 3.18 and 3.19 it can be seen that with a relatively small source-receiver distance, say less than 5–10 m, the RT30/EDT ratio is greater than 1 and the decay curves are concave, especially the early part. An important reason for this is that in a certain period after the direct sound, most first reflections cannot reach a receiver in the near field due to their long sound paths in comparison with the source-receiver distance and, thus, at the receiver there is a rapid decrease in sound energy after the direct sound. With the increase of source-receiver distance, the difference in sound path length between direct sound and reflections becomes less. Consequently, due to the substantial increase in initial energy, the decay curves become convex and the EDT becomes greater than RT30, which can be seen in Figures 3.18

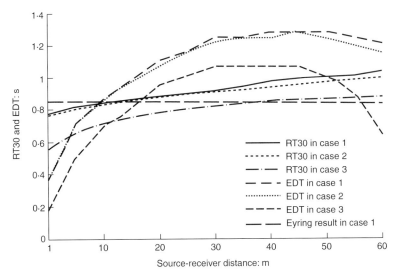

Figure 3.18. Reverberation times in the three cases in Figure 3.17

and 3.19. With further increase of the source-receiver distance, the initial
energy reduces due to the increasing sound path length, and this reduction is
relatively greater than that in later energy since the increase in sound path
length for later energy is relatively less. As a result, the difference between
RT30 and EDT is diminished beyond a certain source-receiver distance.

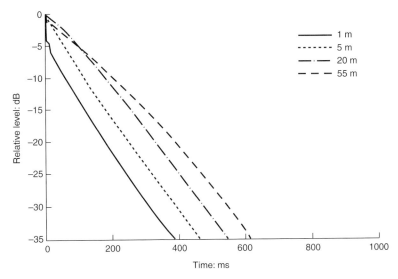

*Figure 3.19. Decay curves at four typical source-receiver distances — calcula-
tion based on case 1 in Figure 3.17*

The result of the Eyring formula is also shown in Figure 3.18. Although it is approximately in the same range as the RT30 calculated by the radiosity method, their tendencies along the length are fundamentally different. This means that in long enclosures with diffusely reflecting boundaries the sound field is not necessarily diffuse.

To investigate the effect of end walls, calculation is made with both ends open, namely $\alpha_{U,V} = 1$, as illustrated in case 2 of Figure 3.17. The calculated RT30 and EDT are shown in Figure 3.18. It can be seen that by opening the ends, the RT30 and EDT are only slightly reduced. This is different from the situation with geometrically reflecting boundaries, where strongly reflecting end walls can significantly increase the reverberation. An important reason for this difference is that with geometrically reflecting boundaries, opening the end walls will make the reverberation at a receiver dependent only on the boundaries in the source-receiver range rather than on all the boundaries, whereas, with diffusely reflecting boundaries, the end walls do not have this effect. As expected, the reduction in reverberation caused by opening the end walls becomes greater the closer a receiver is to the end wall on the receiver side, which can be seen in Figure 3.18.

Calculation is also carried out by 'removing' the boundaries between $x = 0$–29.5 m and $x = 90.5$–120 m. In other words, the length is reduced to 61 m, as illustrated in case 3 of Figure 3.17. The end walls are still open. In Figure 3.18, the RT30 and EDT in this case are also shown. It is seen that the RT30 and EDT are significantly reduced, typically by 20–30%. This suggests that the removed boundaries have considerable effects on the relatively late part of sound decay.

A comparison of RT30 and EDT between diffusely and geometrically reflecting boundaries is shown in Figure 3.20. The comparison is based on case 3 in Figure 3.17. From Figure 3.20 it is seen that in comparison with geometrically reflecting boundaries, the RT30 and EDT with diffusely reflecting boundaries are generally longer, except in the near field. The difference is typically 30–60%. A possible reason for the difference is that with diffusely reflecting boundaries the sound path is generally longer. From Figure 3.20 it can also be seen that with geometrically reflecting boundaries the RT30/EDT ratio is greater than 1, whereas with diffusely reflecting boundaries this ratio is generally less than 1 beyond a certain source-receiver distance. This is probably because the number of reflections with diffusely reflecting boundaries becomes greater with increasing source-receiver distance than with geometrically reflecting boundaries and, consequently, the decay becomes faster with a larger source-receiver distance.

It is noted that in the above comparison between the two kinds of boundary, the absorption is evenly distributed in cross-section, no air absorption is considered, and the two end walls are totally absorptive. If one boundary is strongly absorbent, for example, in a street canyon, the reverberation with

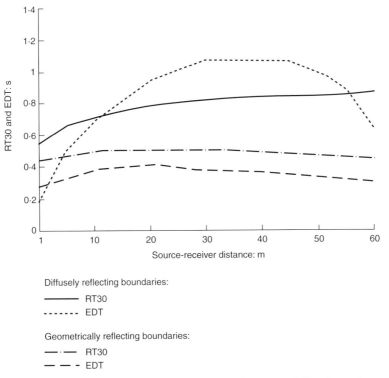

Figure 3.20. Comparison of reverberation time between diffusely and geometri-cally reflecting boundaries — calculation based on case 3 in Figure 3.17

geometrically reflecting boundaries could be longer than that with diffusely reflecting boundaries. The reverberation in street canyons is discussed in Chapter 4. Moreover, if air absorption is included, the difference in reverberation between the two kinds of boundary can be diminished, as analysed in Section 3.3.2. Furthermore, strongly reflective end walls could bring an increase in reverberation, and this increase is greater if the boundaries are geometrically rather than diffusely reflective.

Figure 3.21 shows the sound attenuation along the length with reference to the SPL at 1 m from the source for the three cases in Figure 3.17. It can be seen that the variation in SPL between the three cases is much less than that in reverberation. The main reason, as mentioned previously, is that reverberation is dependent on multiple reflections, whereas the SPL depends mainly on early reflections.

3.3.2.   Air absorption
Since the radiosity model is mainly applicable to relatively high frequencies, it is useful to investigate the effect of air absorption. For case 1 in Figure 3.17, Figure 3.22 compares the RT30 and EDT without air absorption and with

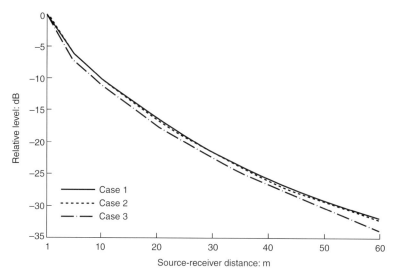

Figure 3.21. Relative SPL attenuation along the length in the three cases in Figure 3.17

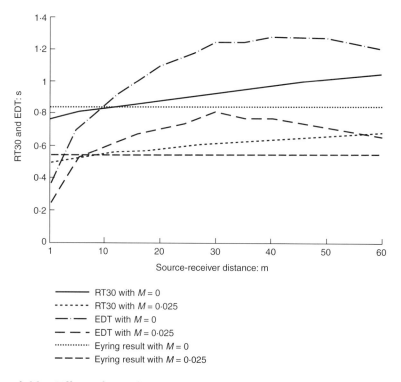

Figure 3.22. Effect of air absorption on reverberation — calculation based on case 1 in Figure 3.17

an air absorption of $M = 0.025\,\text{Np/m}$, which is approximately the 8000 Hz value at a temperature of 20°C and relative humidity of 40–50% [3.6]. As expected, with air absorption the RT30 and EDT decrease considerably. In Figure 3.22 the typical decrease is about 30–45%. It is interesting to note that this ratio of decrease is approximately the same as the calculation by the Eyring formula, which is also shown in Figure 3.22.

For the same configuration, with geometrically reflecting boundaries the decrease in reverberation caused by the air absorption is around 10%, which is much less than that with diffusely reflecting boundaries. An important reason for the difference is that with diffusely reflecting boundaries the sound path is generally longer and, thus, air absorption is more effective.

Calculation has also shown that with diffusely reflecting boundaries the extra SPL attenuation along the length caused by a given air absorption is greater than with geometrically reflecting boundaries. In the above case, for example, with diffusely reflecting boundaries the extra SPL attenuation caused by the air absorption is 3–8·3 dB at 10–60 m from the source, whereas with geometrically reflecting boundaries this extra attenuation is 2·1–7·3 dB.

3.3.3. Cross-sectional area and aspect ratio

Corresponding to Figure 3.12, calculation is carried out with five cross-sections, namely 12 m by 8 m, 16 m by 6 m, 24 m by 4 m, 6 m by 4 m and 8 m by 8 m. In the calculation $L = 120\,\text{m}$, $\alpha_{\text{F,C,A,B,U,V}} = 0.2$, and the source (at $x = 30\,\text{m}$) and receivers are along the centre of the cross-section. The calculated RT30 and EDT are shown in Figure 3.23.

With a constant aspect ratio, as expected, the RT30 and EDT are longer for a larger cross-section, which can be seen by comparing cross-sections 6 m by 4 m and 12 m by 8 m. With a greater cross-section, the RT30/EDT ratio becomes greater in the near field and the point where RT30 = EDT shifts further from the source. This is in correspondence with the fact that the effect caused by increasing cross-sectional area is similar to that of decreasing source-receiver distance.

For a given cross-sectional area, the reverberation could vary significantly with the aspect ratio. By comparing cross-sections 16 m by 6 m, 12 m by 8 m and 24 m by 4 m, it appears that the RT30 and EDT become greater as the cross-section tends towards square. This is probably because the average sound path length reaches a maximum as the aspect ratio tends to 1. For this reason, the reverberation may be longer with a smaller cross-sectional area but a square cross-section (compare 8 m by 8 m and 24 m by 4 m). Similar results have also been obtained for geometrically reflecting boundaries, as described in Section 3.2.3.

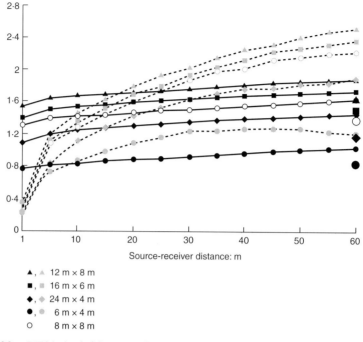

Figure 3.23. RT30 (solid lines with solid symbols) and EDT (dotted lines with grey symbols) with various cross-sections — the corresponding Eyring results are shown on the right side of the figure with large symbols

In Figure 3.23 the results using the Eyring formula are also shown. It is interesting to note that for both the RT30 and EDT, the ratios between various cross-sections are approximately the same as the Eying results.

3.3.4. Distribution of boundary absorption

In order to investigate the variation in reverberation caused by the location of absorption in cross-section, a comparison based on case 1 in Figure 3.17 is made with five different distributions of a constant amount of absorption: case D1, $\alpha_{C,F,A,B} = 0.2$; case D2, $\alpha_{C,F} = 0.32$ and $\alpha_{A,B} = 0.02$; case D3, $\alpha_{F,A} = 0.38$ and $\alpha_{C,B} = 0.02$; case D4, $\alpha_C = 0.62$ and $\alpha_{F,A,B} = 0.02$; and case D5, $\alpha_A = 0.92$ and $\alpha_{C,F,B} = 0.02$. The calculated RT30 and EDT in the five cases are shown in Figure 3.24. It is important to note that from case D1 to D5 the RT30 and EDT decrease continuously and the variation is about 10–30%. In other words, the RT30 and EDT are the longest with absorption that is evenly distributed in cross-section and the shortest when one boundary is strongly absorbent. A simplified model for explaining the differences is that, after hitting each boundary once, the energy loss of a reflection generally increases from case D1 to D5. Assume that the

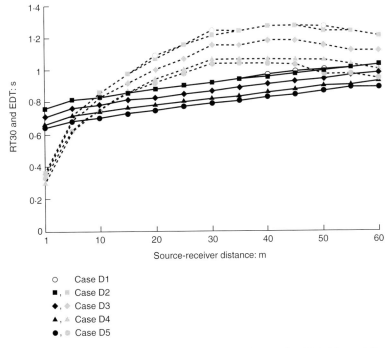

RT30 and EDT: s

Source-receiver distance: m

○	Case D1
■, ▪	Case D2
◆, ◇	Case D3
▲, △	Case D4
●, ○	Case D5

Figure 3.24. RT30 (solid lines with black symbols) and EDT (dotted lines with grey symbols) with various distributions of absorption

original energy of a reflection is 1, the remaining energy is then $0.8^4 = 0.41$ in case D1, $0.68^2 \times 0.98^2 = 0.44$ in case D2, $0.62^2 \times 0.98^2 = 0.37$ in case D3, $0.38 \times 0.98^3 = 0.36$ in case D4 and $0.08 \times 0.98^3 = 0.08$ in case D5.

3.4. Multiple sources

This Section analyses the sound behaviour in long enclosures with multiple sources, including reverberation, sound distribution and the STI. By using MUL (Section 2.6.4), two long enclosures are analysed [3.7]. One is an imaginary rectangular enclosure with geometrically reflecting boundaries of a uniform absorption coefficient α. The cross-section is 6 m by 4 m and the length is 120 m. The other is an underground station of circular cross-sectional shape (London St John's Wood). The diameter and length of this station are 6·5 m and 128 m respectively. The boundary conditions along the station are approximately the same. The sources are point sources in the rectangular enclosure and loudspeakers with the same directionality in the station. In both enclosures it is assumed that the end walls are totally absorbent and there is neither time delay nor sound power difference between the sources, i.e. $t_j = 0$

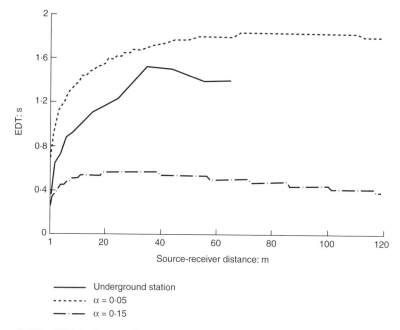

Figure 3.25. EDT of a single source in the underground station and the rectangular enclosure with α = 0·05 and 0·15

and $\Delta S_j = 0$. In the following, for the sake of convenience, the two enclosures are called 'the rectangular enclosure' and 'the underground station', respectively.

3.4.1.   Reverberation from multiple sources

Figure 3.25 shows the EDT of a single source in the above two enclosures. For the rectangular enclosure the EDT is the calculated value with α = 0·05 and 0·15. For the underground station the EDT is the measured value at 630 Hz (octave) in a 1:16 scale model (see Section 5.2), where the source was a tweeter (Foster Type E120T06). As expected, in both enclosures the EDT varies systematically along the length.

To demonstrate the difference in reverberation between a single source and multiple sources, a series of comparisons is made using equation (2.113) for the rectangular enclosure. The configurations are shown in Figure 3.26. It is assumed that the multiple sources are evenly distributed along the whole length with a source spacing σ. Two source spacings, $\sigma = 5$ m and 15 m, are considered. The corresponding source numbers are 25 and 9, respectively. The receivers are at two typical positions: R_0 in the same cross-section with the central source (60 m to both end walls) and R_m with a distance of $\sigma/2$ from the cross-section with the central source. Corresponding with the calculations of a single source in Figure 3.25, α = 0·05 and 0·15 are considered. The

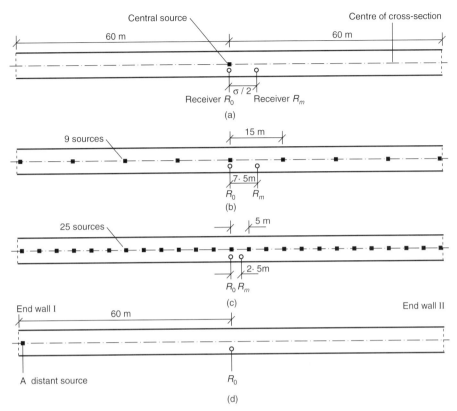

Figure 3.26. Diagrams of the rectangular long enclosure (120 m by 6 m by 4 m) — plan or length-wise section: (a) the central source alone, the source is 2 m from R_0; (b) 9 sources; (c) 25 sources; (d) a source with a distance of 60 m from R_0

comparisons of $L(t)$ curves between a single source and multiple sources at R_0 and R_m are shown in Figures 3.27 and 3.28.

It is interesting to note in Figures 3.27 and 3.28 that the reverberation from multiple sources is significantly longer than that caused by the central source alone. In other words, the reverberation time can be increased by adding sources beyond a single source (compare Figure 3.26(a) with Figures 3.26(b) or 3.26(c)). For example, at R_m the EDT of $\alpha = 0.05$ is 1.98 s with 25 sources and 1.05 s with the central source alone. The reason for this increase is that the later energy of the decay process is increased by the added sources. The effect of these sources can be clearly observed from the curves in Figures 3.27 and 3.28. Similar curves have also been obtained in measurement [3.8].

In contrast, it can be seen in Figure 3.27 that the reverberation from multiple sources is shorter than that caused by the source at 60 m. At R_0 the RT30

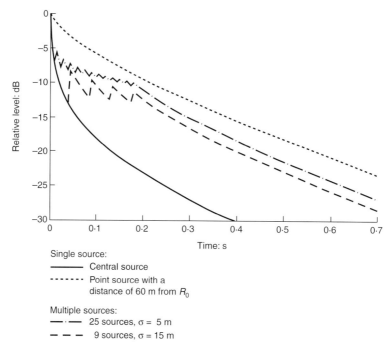

Figure 3.27. L(t) *curves at* R_0 *in the rectangular enclosure with a single source and with multiple sources* — $\alpha = 0.05$

caused by this source is 2·3 s, which is longer than that of 25 sources — 2·17 s; or of 9 sources — 2·14 s. This means that the reverberation time can be decreased by adding sources between a single source and the receiver (compare Figures 3.26(b) or 3.26(c) with Figure 3.26(d)). The decrease is due to the reduction of the minimum source-receiver distance.

The two opposite effects noted above occur simultaneously when the number of evenly distributed sources is changed in a given space. At R_m, with more sources the reduction of the minimum source-receiver distance can decrease the reverberation time but, conversely, the increase of the source number beyond the nearest source may increase the reverberation time. At R_0, the reverberation time should increase with increasing source number since the minimum source-receiver distance is constant. However, this increase might be prohibited, since the initial energy of the decay process could be increased by the added sources that are relatively close to R_0. Overall, because the variation in reverberation and SPL attenuation along the length is complicated, it appears that there is no simple relationship between the reverberation time from multiple sources and the source spacing.

The above analyses demonstrate that the concept of reverberation in long enclosures is fundamentally different from that of the diffuse field: it is dependent on the source position and the number of sources.

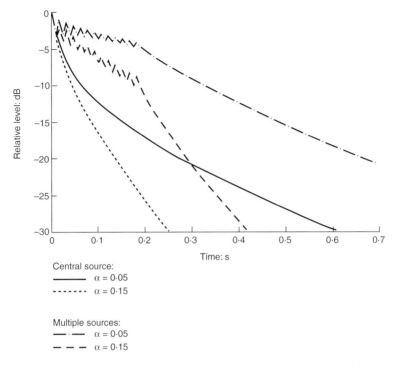

Figure 3.28. L(t) curves at R$_m$ in the rectangular enclosure with the central source and multiple sources of σ = 5 m

Site measurements in a Hong Kong MTR station show that, compared with the EDT (1000 Hz) from multiple loudspeakers, 3·9 s, the EDT of a single loudspeaker is shorter within 30 m, 1·9–3·8 s, but longer beyond 30 m, 4–4·3 s [3.9]. This is in correspondence with the theoretical analysis above.

3.4.2.   Reduction of reverberation from multiple sources

The reduction of reverberation is very useful for reducing noise in the case of multiple noise sources and for increasing the STI in the case of multiple loudspeaker PA systems.

An effective way to reduce the reverberation from multiple sources is to increase the absorption. The effectiveness of this method can be seen in Figure 3.28. However, in the case of multiple loudspeaker PA systems, with more absorption the overall SPL is lower and, thus, the STI could be reduced due to the decrease of the S/N ratio. Moreover, the sound distribution, and correspondingly the STI in the case of high S/N ratio, are more uneven along the length. These negative effects caused by more absorption are shown in Figure 3.29, which corresponds with Figure 3.28. Furthermore, with more absorption the peaks and troughs on L(t) curves are more

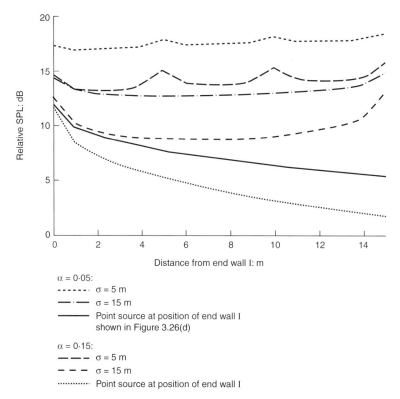

Figure 3.29. Sound distribution in the rectangular enclosure with $\alpha = 0.05$ and $\alpha = 0.15$

significant, which might be subjectively uncomfortable. As a result, for a multiple loudspeaker PA system, the amount of absorption should be based principally on the STI.

   Another possible way to reduce the reverberation from multiple sources is to use directional sources, as mentioned in Section 3.2.4. When the radiation intensity of the sources in the direction of the cross-section is stronger than that in the direction of the length, the sound attenuation along the length can be increased and the effect from further sources can, therefore, be reduced. The negative effect, however, should also be considered: the reverberation time caused by a single source might be increased since the radiation intensity for later reflections is relatively increased. These opposite effects can be seen by comparing the $L(t)$ curves from multiple point and directional sources in the rectangular enclosure, as shown in Figure 3.30, where the directional sources are assumed as $f(\beta) = \sin^3(\beta)$, and the intensity radiation is given in Figure 3.31. It can be demonstrated that with the directional sources the reverberation time is shorter at R_0 but longer at R_m. Moreover, with these directional sources the sound distribution along the length is more uneven.

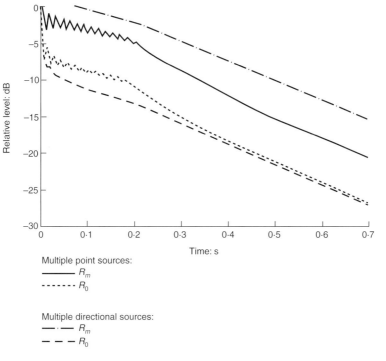

Figure 3.30. L(t) curves in the rectangular enclosure with multiple point and directional sources — σ = 5 m and α = 0·05

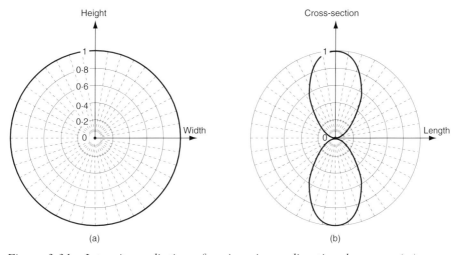

Figure 3.31. Intensity radiation of an imaginary directional source: (a) cross-section; (b) length-wise section

To avoid the above negative effects, directional sources should be so designed that the radiation intensity is the same in the range of $\pm\sigma/2$ and decreases significantly beyond this range. In this case, the reverberation time from multiple sources is close to that of the nearest source alone, which is much shorter.

3.4.3. Speech intelligibility

To demonstrate the comprehensive effect of both reverberation and the S/N ratio on the STI from multiple loudspeakers, calculations are carried out using MUL in the underground station. The configuration is illustrated in Figure 3.32. In the calculation:

(a)   similar to the rectangular enclosure, two typical receivers are chosen, R_0 and R_m;
(b)   two background noise levels are considered, namely 67 dB(A) and 40 dB(A);
(c)   the signal level at R_0 from the central loudspeaker is 80 dB(A);
(d) $\sigma = 4\text{--}14\,\text{m}$; and
(e) $L(t)_z$ is the measured values in the scale model.

The calculation results are shown in Figure 3.33. When the background noise is 40 dB(A), both at R_0 and R_m, the STI increases with the increase of loudspeaker spacing σ. This corresponds to the decrease in reverberation time, which means that in this case the reverberation plays a more dominant role than the S/N ratio. When the background noise is 67 dB(A), at R_0 the STI is systematically lower than that with a background noise of 40 dB(A). It is noted, however, that the STI increases by lengthening the loudspeaker spacing, which indicates that the reverberation is still more dominant than the S/N ratio. Conversely, at R_m, with increasing loudspeaker spacing the STI generally decreases and indicates a minimum. In this case the S/N ratio becomes a significant consideration.

The above analysis demonstrates that with the provision of room and loudspeaker conditions, an optimal loudspeaker spacing can be determined

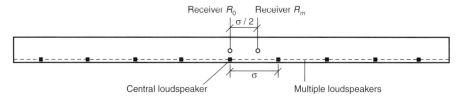

Figure 3.32. Calculation configuration in the underground station (128 m long and 6·5 m in diameter) — plan view

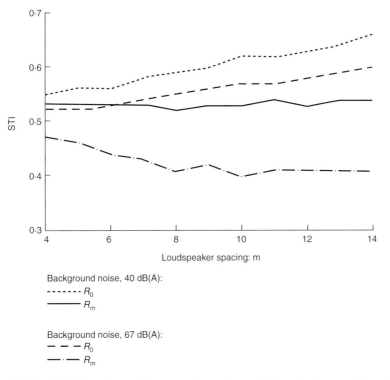

Figure 3.33. STI in the underground station with a background noise of 40 dB(A) and 67 dB(A)

by considering various factors. These include the minimum requirement of the STI, the distribution of the STI along the length and the cost of the whole PA system. For determining such a loudspeaker spacing, the overall prediction method described in Section 2.6 can be used.

The calculation by MUL has been validated by the measurement in the scale model of the underground station. By inputting the measured data of a single loudspeaker, the calculated STI from multiple loudspeakers with $\sigma = 4\,\text{m}$ showed good agreement with measurements. Under two conditions in the scale model, namely with and without suspended absorbers along the ceiling, the difference in the STI from multiple loudspeakers between calculation and measurement [3.10] was within ± 0.03, both at R_0 and R_m.

3.5. Summary

To provide a concise design guidance, the results in this Chapter are summarised below. Although the calculations are mainly for rectangular long enclosures, the principles should also be applicable for long enclosures with other cross-sectional shapes.

3.5.1. Sound attenuation along the length

(a) In the case of geometrically reflecting boundaries, with a larger cross-sectional area the relative attenuation with reference to a given distance from the source becomes less but the absolute attenuation with reference to the sound power level of the source is greater. In the case of diffusely reflecting boundaries, both kinds of attenuation become less with a larger cross-sectional area.

(b) With a constant cross-sectional area, when the width/height ratio is greater than 1 (say 4:1) the sound attenuation tends to become slightly greater than that of the square section.

(c) The efficiency of absorbers per unit area is higher when there is less absorption.

(d) With a given amount of absorption, if the boundaries are geometrically reflective, absorbers should be evenly arranged in cross-section in order to obtain a higher attenuation. With diffusely reflecting boundaries, the difference in sound attenuation between various absorber distributions is not significant.

(e) For diffusely, as opposed to geometrically, reflecting boundaries, the sound attenuation along the length is considerably greater.

3.5.2. Reverberation

(a) With geometrically reflecting boundaries, when the source-receiver distance increases both the RT30 and EDT increase rapidly to a maximum and then decrease slowly. With diffusely reflecting boundaries, the EDT shows a similar tendency but the RT30 tends to increase continuously with increasing source-receiver distance.

(b) The decay curves are not linear. With geometrically reflecting boundaries the decay curves are concave, whereas with diffusely reflecting boundaries the decay curves are concave in the near field and then become convex.

(c) With a constant cross-sectional area, the reverberation time reaches a maximum as the aspect ratio tends to 1.

(d) With a constant aspect ratio, both the RT30 and EDT increase with increasing cross-sectional area and this increase is generally more significant for a longer source-receiver distance.

(e) For diffusely, as opposed to geometrically, reflecting boundaries, the air absorption is more effective with regard to both reverberation and sound attenuation.

(f) Highly reflective end walls can increase the reverberation time and this is more significant for geometrically reflecting boundaries than for diffusely reflecting boundaries.

(g) For diffusely reflecting boundaries, with a given amount of absorption, the RT30 and EDT are the longest with absorption that is evenly distributed in cross-section and the shortest when one boundary is strongly absorbent.

(*h*) With geometrically reflecting boundaries the reverberation time can be varied using a directional source.

3.5.3. Multiple sources

(*a*) The reverberation time at a receiver can be increased significantly by adding sources beyond a single source and, conversely, the reverberation time can be decreased by adding sources between a single source and the receiver.

(*b*) The reverberation from multiple sources can be reduced by more absorption or by carefully designed directional sources. However, with more absorption the STI of a multiple loudspeaker PA system is not necessarily better.

(*c*) For a multiple loudspeaker PA system, depending on room and loudspeaker conditions, there is an optimal loudspeaker spacing for the STI.

3.6. References

3.1   KANG J. Sound attenuation in long enclosures. *Building and Environment*, 1996, **31**, 245–253.

3.2   KANG J. and LU X. Principles and applications of suspended mineral wool absorbers. *Proceedings of the Acoustical Society of Beijing*, 1992, 188–198 (in Chinese).

3.3   KANG J. Reverberation in rectangular long enclosures with geometrically reflecting boundaries. *Acustica/Acta Acustica*, 1996, **82**, 509–516.

3.4   KANG J. *Acoustics of long enclosures*. PhD Dissertation, University of Cambridge, England, 1996.

3.5   KANG J. Reverberation in rectangular long enclosures with diffusely reflecting boundaries. *Acustica/Acta Acustica*, 1999, **85**, S389–390.

3.6   AMERICAN NATIONAL STANDARDS INSTITUTE (ANSI). *Method for the calculation of the absorption of sound by the atmosphere*. ANSI S1.26. ANSI, 1995 (Revised 1999).

3.7   KANG J. Acoustics in long enclosures with multiple sources. *Journal of the Acoustical Society of America*, 1996, **99**, 985–989.

3.8   ORLOWSKI R. J. *London Underground Ltd Research — Acoustic control systems for LUL cut and cover stations*. Arup Acoustics Report, No. AAc/46318/02/01, 1994.

3.9   KANG J. *Station acoustics study*. Hong Kong Mass Transit Railway Corporation Report, Consultancy No. 92110800-94E, 1995/96.

3.10   ORLOWSKI R. J. *London Underground Ltd Research — Acoustic control systems for deep tube tunnels*. Arup Acoustics Report, No. AAc/46318/01/03, 1994.

4. Parametric study: design guidance for urban streets

This chapter presents a series of parametric studies for urban streets. The objective is to analyse basic characteristics of the sound fields in urban streets and to demonstrate the effectiveness of architectural changes and urban design options on noise reduction. Major parameters considered include street aspect ratio, boundary absorption, strategic absorber arrangement, absorption from air and vegetation, source number and position, and reflection characteristics of boundaries. The analysis is focused on sound distribution, especially the sound attenuation along the length, but reverberation is also considered. The chapter begins with an analysis of the sound field in a single urban street with diffusely reflecting boundaries, followed by a comparison of sound fields resulting from diffusely and geometrically reflecting boundaries. It then analyses the sound field in interconnected streets.

The radiosity model is used for diffusely reflecting boundaries and the image source method is applied to geometrically reflecting boundaries. For the sake of convenience, the boundary absorption is assumed to be independent of the angle of incidence. Excess attenuation due to ground interference and temperature or wind-gradient induced refraction is not taken into account. Except where indicated, absorption from air and vegetation is not included, and the SPL at receivers is relative to the source power level, which is set as zero.

4.1. A single street with diffusely reflecting boundaries

4.1.1. Sound distribution in a typical urban street

To investigate basic characteristics of the sound field resulting from diffusely reflecting boundaries, a typical street configuration is considered, as illustrated in Figure 4.1. The street length, width and height are 120 m, 20 m and 18 m, respectively. The buildings are continuous along the street and of a constant height on both sides. A point source is positioned at (30 m, 6 m and 1 m). The façades and ground have a uniform absorption coefficient of 0·1. The patch

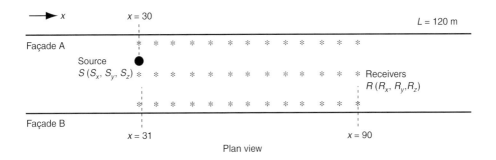

Plan view

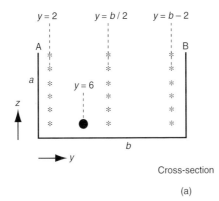

Cross-section

(a)

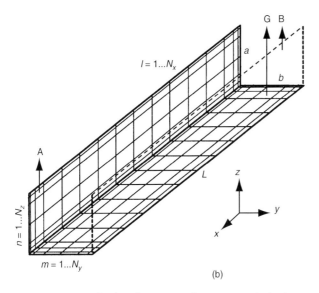

(b)

Figure 4.1. An idealised rectangular street: (a) plan and cross-section showing the source and receiver positions used in the calculation (dimensions in metres); (b) three-dimensional projection of the street showing an example of patch division

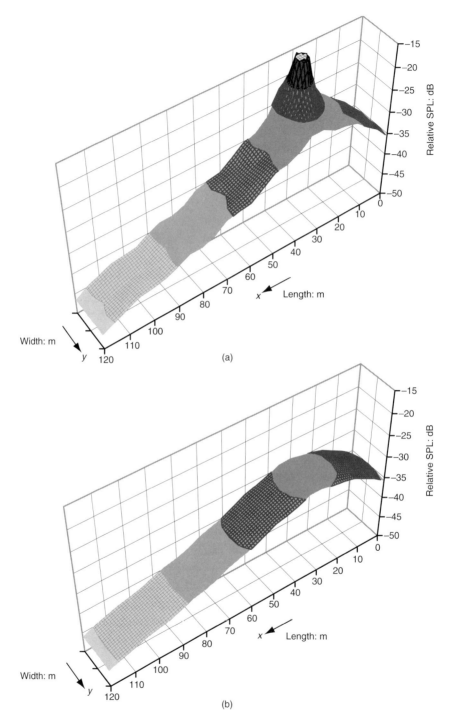

Figure 4.2. SPL distribution on two horizontal planes in a typical street: (a) 1 m above the ground; (b) 18 m above the ground — each colour represents 5 dB

division is made with $N_X = 60$ and $N_Y = N_Z = 8$. Along the length the patch size increases from $l = 1$ to 15, decreases from $l = 46$ to 60 and is constant between $l = 16$ and 45. For the varied patch sizes, the ratio between two adjacent patches is $q_x = 1.2$. Along the width and height, the patch size increases from the edges to the centre with a ratio of $q_y = q_z = 1.5$. Using these parameters, the program calculates the form factors and the source energy distribution to first-order patch sources accurate to four decimal places.

The SPL distribution on a horizontal plane at 1 m above the ground is shown in Figure 4.2(a). It can be seen that although the boundaries are diffusely reflective, the SPL varies significantly on the plane. Along $y = 10$ m, for example, the SPL attenuation is 22·7 dB at source-receiver distances of 5 m through to 90 m (i.e. 26·7 dB/100 m). As expected, the SPL variation becomes less when the horizontal plane is farther from the source. In Figure 4.2(b) the sound distribution on a plane of 18 m above the ground is shown. Along $y = 10$ m the SPL attenuation at source-receiver distances of 5 m through to 90 m becomes 16·8 dB (i.e. 19·8 dB/100 m). By comparing Figures 4.2(a) and 4.2(b), it can be seen that the SPL difference between the two planes decreases with increasing distance from the source in the length direction. Beyond the source-receiver distance of 10–15 m, this difference is less than 1–2 dB. This indicates that in this range the sound distribution in cross-section is rather even.

The sound distribution on a vertical plane at a distance of 1 m from façade A is shown in Figure 4.3(a). At $x = 30$ m, namely the source position, the SPL variation with height is significant, at about 8 dB. With increasing distance from the source in the length direction, the SPL variation with height

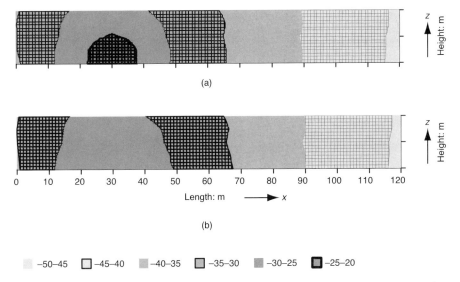

Figure 4.3. Relative SPL on two vertical planes in a typical street: (a) 1 m from façade A; (b) 1 m from façade B

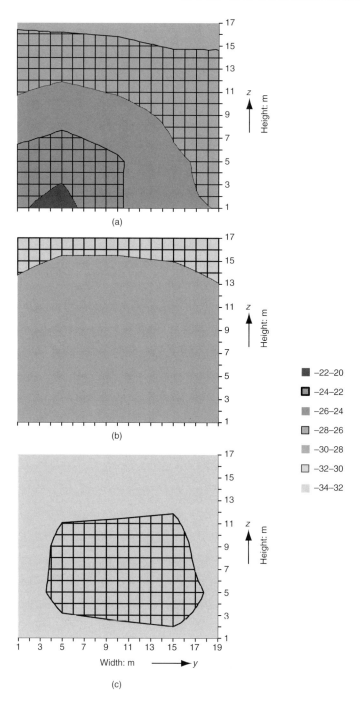

Figure 4.4. Relative SPL in three cross-sections in a typical street: (a) 5 m (x = 35 m) from the source; (b) 15 m (x = 45 m) from the source; (c) 25 m (x = 55 m) from the source

decreases rapidly. Beyond $x = 50$ m, it becomes less than 1 dB. In Figure 4.3(b) the sound distribution on a vertical plane which is 1 m from façade B is shown. It can be seen that the SPL variation with height is generally less than that in Figure 4.3(a). At $x = 30$ m, for example, the variation is only 2·5 dB. Clearly, this is because façade B is farther from the source than façade A.

The sound distribution in three typical cross-sections at $x = 35$ m, 45 m and 55 m is shown in Figure 4.4. At $x = 35$ m the SPL variation in the cross-section is 7·5 dB and with increasing source-receiver distance in the length direction this variation decreases rapidly. Beyond $x = 55$ m, the variation is less than 1·5 dB.

4.1.2. Street aspect ratio

To investigate the effect of street aspect ratio on the sound field, a range of street heights from 6 m to 54 m is considered, which corresponds to the width/height ratio of 3·3 to 0·37. Other configurations are the same as those in Figures 4.1 to 4.4. Figure 4.5 shows a comparison of sound attenuation between four street heights, $a = 6$ m, 18 m, 30 m and 54 m, where the receivers are at (31–90 m, 2 m, 1 m) and the sound levels are with reference to the SPL at 1 m from the source when $a = 54$ m. It can be seen that the SPL attenuation along the length becomes less with increasing street height. An apparent reason is that with a greater street height, less energy can be reflected out of the street canyon. From Figure 4.5 it is also seen that the SPL attenuation curve along the length is concave. In other words, the attenuation per unit distance becomes less with the increase of source-receiver distance. The attenuation

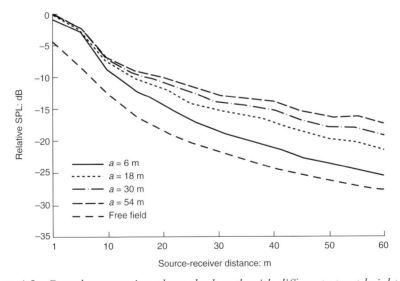

Figure 4.5. Sound attenuation along the length with different street heights

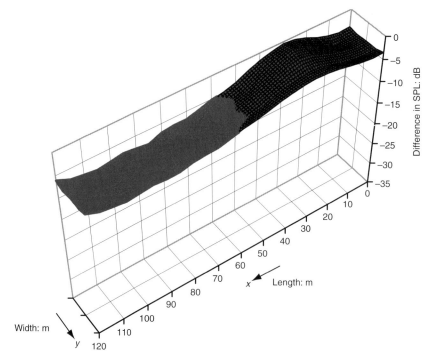

Figure 4.6. Extra SPL attenuation on a horizontal plane at 1 m above the ground caused by reducing the street height from 54 m to 6 m

curve in the free field is also shown in Figure 4.5. It is interesting to note that the difference between $a = 6$ m and the free field is only about 2–4 dB.

Figure 4.6 shows the SPL difference between street heights 6 m and 54 m, where the receivers are on a horizontal plane at a distance of 1 m from the ground. In the near field, say within 10 m from the source, the difference between the two street heights is insignificant, which indicates the strong influence of the direct sound. With the increase of source-receiver distance, the effect of boundaries becomes more important and, thus, the difference between the two street heights becomes greater. Beyond about $x = 110$ m, there is a decrease in the SPL difference. This is because there is no boundary beyond $x = 120$ m and, consequently, the boundary effect is diminished.

It is noteworthy that with $a = 54$ m, although the street width/height ratio is rather small, the SPL still varies significantly along the length. Along $y = 10$ m, for example, the SPL attenuation at source-receiver distances of 5 m through to 90 m is 19·3 dB (i.e. 22·7 dB/100 m).

4.1.3. Boundary absorption and building arrangements
Boundary absorption is useful for diminishing reflection energy and, consequently, reducing the overall SPL in urban streets. Broadly speaking,

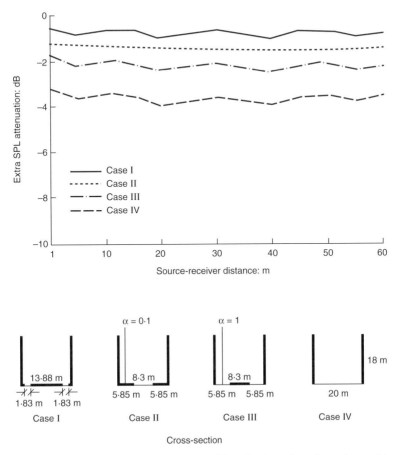

Figure 4.7. Extra SPL attenuation caused by placing absorbers ($\alpha = 1$) on the ground

absorbers on street boundaries include absorbent materials, open windows as sound energy sinks and gaps between buildings. This Section analyses the effectiveness of boundary absorption and strategic building arrangements upon the sound attenuation along the length. For the sake of convenience, the location of absorbers and building blocks corresponds to the patch division in Section 4.1.1. The calculation of the sound attenuation along the length is based on the average of four receiver lines, namely (31–90 m, 2 m, 1 m), (31–90 m, 2 m, 18 m), (31–90 m, 18 m, 1 m), and (31–90 m, 18 m, 18 m), as illustrated in Figure 4.1. The effectiveness of an acoustic treatment, such as adding absorbers, is evaluated by the extra SPL attenuation caused by the treatment with reference to the typical case described in Section 4.1.1.

   Figure 4.7 shows the effect of absorbers on the ground. The absorbers are arranged along the length, and four cases are considered:

(*a*)   case I, absorbers from $y = 1.23$ m to 3.06 m and from $y = 16.94$ m to 18.77 m;

(*b*)   case II, absorbers from $y = 5.85$ m to 14.15 m;

(*c*)   case III, absorbers from $y = 0$ to 5.85 m and from $y = 14.15$ m to 20 m; and

(*d*)   case IV, absorbers over the entire ground.

The percentages of the ground absorption cover in the four cases are 18.3%, 41.5%, 58.5% and 100%, respectively. For convenience, it is assumed that the absorption coefficient of the absorbers is 1. In Figure 4.7 it can be seen

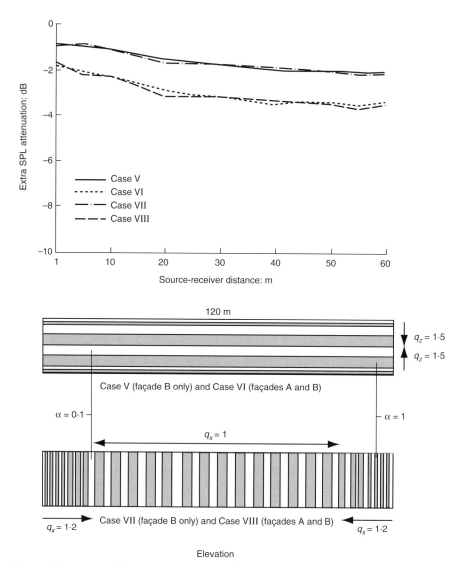

Figure 4.8.   Extra SPL attenuation caused by placing absorbers ($\alpha = 1$) on façades

that from case I to IV the extra attenuation increases continuously. It appears that the extra attenuation is approximately proportional to the absorber area.

The effect of façade absorption is shown in Figure 4.8. Four cases are considered:

(a) case V, absorbers along the length, façade B only;
(b) case VI, absorbers along the length, façades A and B;
(c) case VII, absorbers along the height, façade B only; and
(d) case VIII, absorbers along the height, façades A and B.

For each treated façade, the ratio of absorber to façade area is 50%. Again, the absorption coefficient of the absorbers is assumed as 1. From Figure 4.8 it can be seen that with absorbers on façade B only, the extra attenuation is about 1–2 dB and, with absorbers on both façades, the extra attenuation is around 2–4 dB. It is interesting to note that with a given absorber area there is almost no difference in sound attenuation between vertically and horizontally distributed absorption. This suggests that with diffusely reflecting boundaries, if a given amount of absorbers is evenly distributed on a boundary, the pattern of the absorber arrangement plays an insignificant role for the sound field.

The extra sound attenuation caused by reducing the height of façade B to 15·25 m (case IX), 9 m (case X), 2·75 m (case XI) and 0 m (case XII) is shown in Figure 4.9. As expected, from case IX to XII the extra attenuation increases continuously. With case XII the influence of façade B upon sound attenuation along the street can be seen clearly.

Figure 4.9 shows that in comparison with the case where the absorption coefficient of all the boundaries is 0·1, a similar extra attenuation, 3–4 dB, can be achieved by increasing the absorption coefficient to 0·5, or by taking one side of the buildings away (i.e. case XII) or by treating the ground as totally absorbent (i.e. case IV). The SPL attenuation in the free field is also shown in Figure 4.9, which indicates the limit in noise reduction by boundary absorption and strategic building arrangements.

It is also interesting to investigate the effect of gaps between buildings. Figure 4.10 shows the extra sound attenuation in two cases, namely, case XIII — one gap on façade B between $x = 48$ m and 72 m — and case XIV — one gap on façade A and one gap on façade B, both between $x = 48$ m and 72 m. Since it has been demonstrated that the effect of reflections from building side walls to a street is generally insignificant (see Section 4.3), in the calculation the absorption coefficient of the gap is assumed as 1. From Figure 4.10 it can be seen that in the length range containing the gap(s) there is a considerable extra SPL attenuation, which is about 2 dB in case XIII and 3 dB in case XIV. Conversely, after the gap(s), say, $x > 75$ m, the extra attenuation becomes systematically less and before the gap(s), say, $x < 45$ m, the extra attenuation is almost unnoticeable. These results suggest that although all patches affect the

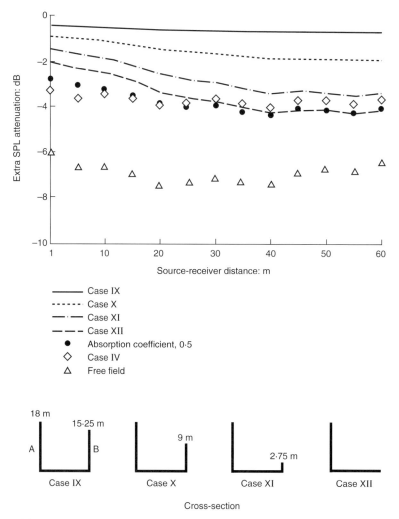

Figure 4.9. Extra SPL attenuation caused by reducing the height of façade B compared with the extra attenuation by increasing the absorption coefficient of all the boundaries to 0·5, and by treating the ground as totally absorbent (case IV) — the SPL attenuation in the free field is also shown

SPL at a receiver because they are diffusely reflective, the patches near the receiver are more effective.

4.1.4.   Reverberation

Reverberation is an important index for the acoustic environment in urban streets. On the one hand, with a constant SPL, noise annoyance is greater with a longer reverberation time [4.1]. On the other hand, a suitable reverberation

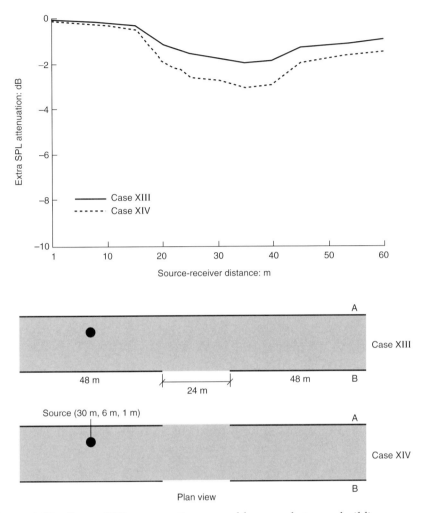

Figure 4.10. Extra SPL attenuation caused by gaps between buildings

time, say, 1–2 s, can make 'street-music' more enjoyable. Reverberation in street canyons is also useful for investigating multiple reflections [4.2–4.4].

Figure 4.11 shows the effect of source-receiver distance, street height and boundary absorption on decay curves. In the calculation the street length and width are 120 m and 20 m, and a point source is positioned at (30 m, 6 m, 1 m). By comparing curves 3–5, where the street height is 18 m but the source-receiver distance varies from 5 m to 60 m, it can be seen that reverberation increases systematically with increasing distance from the source. This is similar to the situation in long enclosures, as described in Chapter 3. By comparing curves 1, 4 and 6, where the source-receiver distance is 20 m but the street height varies from 6 m to 30 m, it can be seen that reverberation increases significantly when the street height becomes greater. The effect of boundary

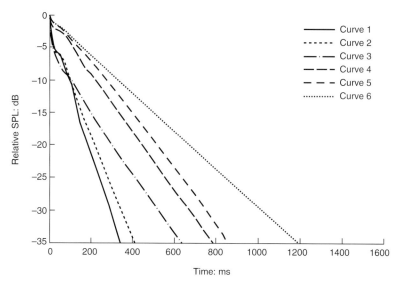

Figure 4.11. Deacy curves in six typical cases: curve 1, street height a = 6 m and source-receiver distance z = 20 m; curve 2, a = 18 m, z = 20 m and the boundary absorption coefficient 0·5; curve 3, a = 18 m and z = 5 m; curve 4, a = 18 m and z = 20 m; curve 5, a = 18 m and z = 60 m; curve 6, a = 30 m and z = 20 m

absorption can be seen by comparing curves 2 and 4, where the reverberation is approximately doubled when the absorption coefficient decreases from 0·5 to 0·1. Overall, in the configurations in Figure 4.11, the RT is about 0·7–2 s. This suggests that the reverberation effect is significant in such a street.

4.2. Comparison between diffusely and geometrically reflecting boundaries

In Chapter 3 it has been demonstrated that in long enclosures there are considerable differences between sound fields resulting from diffusely and geometrically reflecting boundaries. This Section analyses the situation in urban streets [4.5,4.6]. The configuration used in the analysis is generally similar to that illustrated in Figure 4.1. Except where indicated, the street length is 120 m, a point source is positioned at $x = 30$ m and the source-receiver distance along the length is 1–60 m ($x = 31$–90 m). With this arrangement the calculation results should not be affected significantly when the street length is extended. In other words, the results can be generalised to represent a longer street. The façades and ground are assumed to have a uniform absorption coefficient of 0·1. With some strong absorption patches, such as open windows or gaps between buildings as sound energy sinks, it has been demonstrated that the trend of the comparison results will not change systematically.

4.2.1. Sound attenuation along the length

Figure 4.12 shows the extra SPL attenuation along the length caused by replacing geometrically reflecting boundaries with diffusely reflecting boundaries. The calculation is based on two street layouts, namely $b = 20\,\text{m}$ and $a = 6\,\text{m}$, and $b = 20\,\text{m}$ and $a = 18\,\text{m}$. The receivers are along two lines in the length direction, namely (31–90 m, 2 m, 1 m) and (31–90 m, 18 m, a), which represent relatively high and low SPL in a cross-section, respectively. The source is at (30 m, 6 m, 1 m). From Figure 4.12 it is interesting to note that in comparison with geometrical boundaries, the SPL with diffuse boundaries decreases significantly with increasing source-receiver distance. The main reason is that, with diffuse boundaries, the total energy loss becomes greater because the average sound path length is longer, especially for the receivers in the far field. For this reason, if air or vegetation absorption is included, the difference between diffuse and geometrical boundaries is even greater. This is supported by calculation results. In the near field, with diffuse boundaries there is a slight increase in SPL, which is likely due to the energy reflected back from farther boundaries. As expected, this increase becomes less with

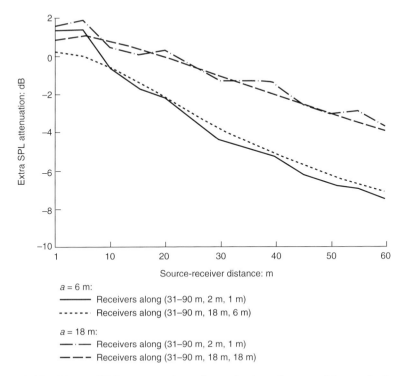

Figure 4.12. Extra SPL attenuation along the length caused by replacing geometrically reflecting boundaries with diffusely reflecting boundaries

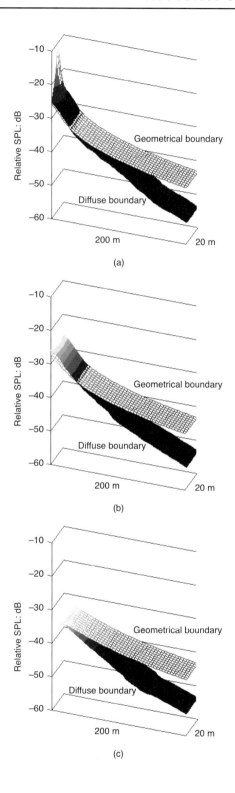

increasing source-receiver distance in the length and/or width direction. The above differences between the two kinds of boundary are similar to those in long enclosures, as described in Chapter 3.

From Figure 4.12 it can also be seen that, with $a = 18$ m, the extra attenuation is less than that with $a = 6$ m, both along (31–90 m, 2 m, 1 m) and (31–90 m, 18 m, a). For receivers along (31–90 m, 2 m, 1 m), an important reason for the difference between $a = 6$ m and 18 m is that, with diffuse boundaries the sound energy increases with increasing street height and this energy increase is proportionally greater for a longer source-receiver distance, whereas with geometrical boundaries the sound attenuation is not affected by the change in street height because the source and receiver heights, S_z and R_z, are less than 6 m. For receivers along (31–90 m, 18 m, a), a notable reason for the difference between $a = 6$ m and 18 m is that in comparison with (31–90 m, 18 m, 6 m), the sound attenuation along (31–90 m, 18 m, 18 m) is considerably less, both with diffuse and geometrical boundaries. Given that along (31–90 m, 18 m, 18 m) the sound attenuation is only 6 dB with geometrical boundaries, the 4 dB extra attenuation caused by replacing geometrical boundaries with diffuse boundaries is significant. If the street width is reduced to 10 m, with geometrical boundaries the sound attenuation along (31–90 m, 8 m, 18 m) increases to 7 dB. Replacing geometrical boundaries with diffuse boundaries, the extra attenuation increases to 5 dB.

Another comparison between the two kinds of boundary is shown in Figure 4.13, where the street length, width and height are 200 m, 20 m and 30 m, respectively. A point source is positioned at (5 m, 10 m, 1 m), namely at one end of the street. Three horizontal receiver planes are considered, which are at 1 m, 10 m and 30 m above the ground. It can be seen that the difference in sound attenuation between diffusely and geometrically reflecting boundaries is significant, typically 10 dB at the far end of the street.

4.2.2. Sound distribution in cross-sections

To investigate the SPL variation in the width and height direction, calculations are performed at 54 receiver locations in a cross-section with a distance of 5 m from the source, for both diffusely and geometrically reflecting boundaries. In the calculation, the cross-section is $b = 20$ m and $a = 18$ m, the source is at (30 m, 6 m, 1 m), and the receivers are along three lines in the height direction with $y = 2$ m, 10 m and 18 m. Figure 4.14 shows the SPL at the receivers with reference to the SPL at (35 m, 2 m, 1 m). For this scenario it can be seen that

Figure 4.13. Comparison of sound distribution between diffusely and geometrically reflecting boundaries: (a) SPL on a horizontal plane at 1 m above the ground; (b) 10 m above the ground; (c) 30 m above the ground

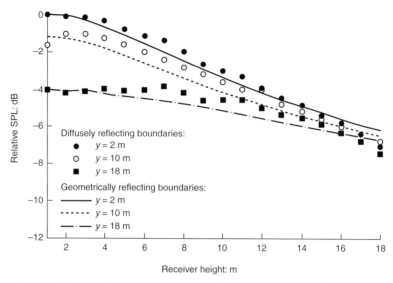

Figure 4.14. SPL distribution along three lines in the height direction

there is no significant difference between diffusely and geometrically reflecting boundaries. For both kinds of boundary the difference between the maximum and minimum SPL in the cross-section is about 7 dB.

As expected, the SPL variation in a cross-section becomes less with increasing distance between the cross-section and the source. For both diffusely and geometrically reflecting boundaries, Figure 4.15 shows the SPL difference between two lines of receiver along the length, (31–90 m, 2 m, 1 m) and (31–90 m, 18 m, a) which represent relatively high and low SPL in a cross-section. It can be seen that when the source-receiver distance is less than about 15 m, the SPL variation is about 2–10 dB. Beyond this range, the variation is generally within 2 dB. With $a = 18$ m the SPL variation is systematically greater than that with $a = 6$ m but the difference is only around 1 dB.

By comparing Figures 4.13(a), 4.13(b) and 4.13(c) it can be demonstrated that, similar to the situation with diffuse boundaries, with geometrical boundaries the SPL variation in a cross-section decreases as the distance between the cross-section and the source increases. With increasing receiver height, as expected, the SPL attenuation becomes less.

The source position may also affect the sound distribution in cross-section. For a street with $b = 20$ m and $a = 18$ m, the decrease in SPL caused by moving a point source from (30 m, 3 m, 1 m) to (30 m, 17 m, 1 m) is shown in Figure 4.16, for both diffusely and geometrically reflecting boundaries, where the receivers are along two lines, namely (31–90 m, 2 m, 1 m) and (31–90 m, 2 m, 18 m). For both kinds of boundary, at receivers (31–90 m, 2 m, 1 m) the SPL difference with the two source positions is significant in the near field but

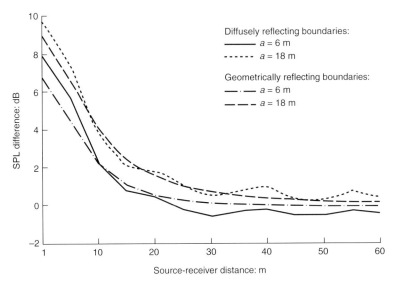

Figure 4.15. SPL difference between two lines of receiver along the length that represents relatively high and low SPL in a cross-section

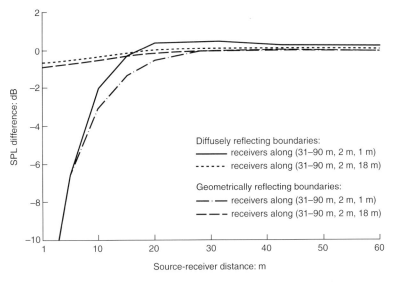

Figure 4.16. SPL difference caused by moving a point source from (30 m, 3 m, 1 m) to (30 m, 17 m, 1 m)

becomes negligible beyond a certain source-receiver distance, say 15 m. At receivers (31–90 m, 2 m, 18 m), the SPL difference caused by moving the source is only within 1 dB.

4.2.3. Street aspect ratio

With geometrically reflecting boundaries the SPL at a receiver is not affected by the street boundaries that are above the source and receiver. In other words, if the source height is lower than the street height, the sound field in the street remains unchanged with further increase of the street height. With diffusely reflecting boundaries, conversely, the SPL at any receiver is affected by the increase of street height. The effect of street height in the case of diffuse boundaries has been analysed in Section 4.1.2.

To investigate the effect of street width, extra SPL attenuation along the length caused by moving façade B from $y = 10$ m to 40 m is calculated with $a = 18$ m, for both diffuse and geometrical boundaries. This corresponds to a change in width/height ratio of 0·56 to 2·2. Figure 4.17 shows the average extra attenuation along two lines of receiver near façade A, namely (31–90 m, 2 m, 1 m) and (31–90 m, 2 m, 18 m), again with a point source at (30 m, 6 m, 1 m). The difference in extra attenuation between the two kinds of boundary is generally insignificant. Interestingly, for both kinds of boundary the variation in the extra SPL attenuation is not monotonic along the length — the maximum occurs at distance from the source of 20 m with diffuse

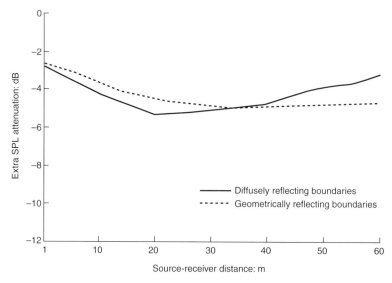

Figure 4.17. Extra attenuation along the length caused by moving façade B from y = 10 m to 40 m

boundaries and of 35 m with geometrical boundaries. This is probably because
in the near field, the SPL is dominated by the initial reflections from façade A
such that façade B is relatively less effective. In the very far field, the average
sound path length is already long and, consequently, the effect caused by
moving façade B away is proportionally less.

Corresponding to Figure 4.13, a comparison of sound distribution between
street widths 5 m and 160 m is shown in Figure 4.18, where the boundaries are
geometrically reflective, the street height is 20 m and the receiver plane is at
10 m above the ground. It can be seen that between the two street widths,

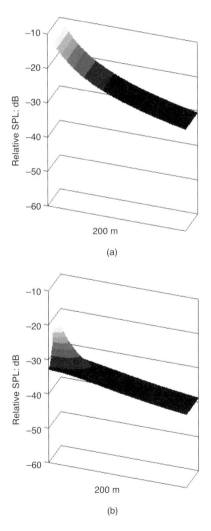

*Figure 4.18. Comparison of SPL distribution between two street widths with
geometrically reflecting boundaries: (a) street width 5 m; (b) street width 160 m*

which correspond to a change in width/height ratio of 0·25 to 8, the SPL difference is about 9 dB at a source-receiver distance of 200 m.

4.2.4. Amount of boundary absorption

For both diffusely and geometrically reflecting boundaries, the extra attenuation caused by evenly increasing the absorption coefficient of all the boundaries from 0·1 to 0·5 and to 0·9 is shown in Figure 4.19, where $b = 20$ m, $a = 18$ m and the source is at (30 m, 6 m, 1 m). The SPL presented is the average of two receiver lines (31–90 m, 2 m, 1 m) and (31–90 m, 18 m, 18 m). It can be seen that with diffusely reflecting boundaries the extra attenuation is almost constant along the street length. This is because in a street with diffusely reflecting boundaries the SPL at any receiver is dependent on the contribution from all the patches and, thus, an even increase of absorption on all the boundaries should have a similar effect on all the receivers. For geometrically reflecting boundaries, the extra attenuation increases with the increase of source-receiver distance. This is possibly because, with a longer source-receiver distance the

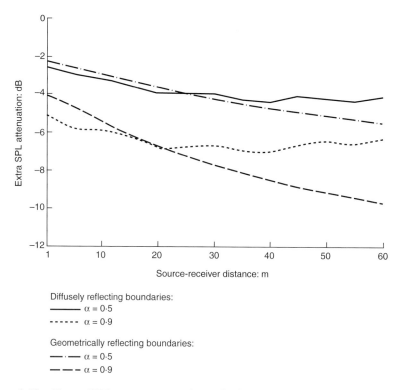

Figure 4.19. Extra SPL attenuation along the length caused by evenly increasing the absorption coefficient of all the boundaries from 0·1 to 0·5 and to 0·9

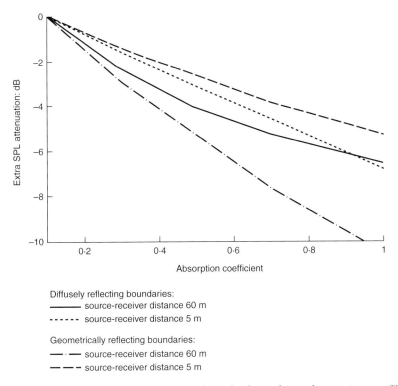

Figure 4.20. Extra SPL attenuation when the boundary absorption coefficient is increased from α = 0·1

difference in sound path length between low and high orders of reflection becomes less, such that higher orders of reflection become relatively important. Given that increasing boundary absorption is more effective for higher orders of reflection, the extra attenuation becomes greater with a longer source-receiver distance.

The increase in SPL attenuation with increasing boundary absorption coefficient is shown in Figure 4.20, where the SPL with α = 0·1 is used as a reference. Two source-receiver distances, 5 m and 60 m, are considered. At each source-receiver distance the SPL is the average of four receivers, namely: (5 m or 60 m, 2 m, 1 m); (5 m or 60 m, 2 m, 18 m); (5 m or 60 m, 18 m, 1 m); and (5 m or 60 m, 18 m, 18 m). It can be seen that the extra attenuation caused by increasing boundary absorption is significant. Generally speaking, for both diffuse and geometrical boundaries, as the absorption coefficient increases linearly, the attenuation increases with a decreasing gradient. This means that with a given increase in absorption, the efficiency in noise reduction is greater when there are fewer absorbers. In long enclosures similar results have also been obtained, as described in Section 3.1.2.

4.2.5. Distribution of boundary absorption

In Section 3.1.3 it has been demonstrated that with a given amount of absorption in a long enclosure, the sound attenuation along the length varies considerably with different distributions of absorption in cross-section. To investigate the situation in urban streets, calculation is made with three distribution schemes of a constant amount of absorption: I, one façade strongly absorbent; II, two façades strongly absorbent; and III, all three boundaries evenly absorbent. The absorption coefficients of the boundaries are: distribution I, $\alpha_A = 0.9$ and $\alpha_B = \alpha_G = 0.05$; distribution II, $\alpha_A = \alpha_B = 0.475$ and $\alpha_G = 0.05$; and distribution III, $\alpha_A = \alpha_B = \alpha_G = 0.209$. In the calculation the cross-section is $b = 20\,\text{m}$ and $a = 6\,\text{m}$, the source is at $(30\,\text{m}, 6\,\text{m}, 1\,\text{m})$ and the SPL in a cross-section is represented by the average of two receivers, namely $(31-60\,\text{m}, 2\,\text{m}, 1\,\text{m})$ and $(31-60\,\text{m}, 18\,\text{m}, 6\,\text{m})$.

Figure 4.21 shows the extra attenuation caused by replacing distribution III with distributions I and II. The extra attenuation with distribution I is typically $1-4\,\text{dB}$ in the case of geometrical boundaries and $1\,\text{dB}$ with diffuse boundaries. With distribution II the extra attenuation is systematically less than that with distribution I but still noticeable. As mentioned in Section 3.3.4, a simplified model for explaining the differences between the different distribution schemes is that, after hitting each boundary once, the remaining energy of a reflection is the lowest with distribution I and the highest with distribution III. Assume that the original energy of a reflection is 1, the remaining energy is then $0.1 \times 0.95^2 = 0.09$ with distribution I, $0.525^2 \times 0.95 = 0.26$ with distribution

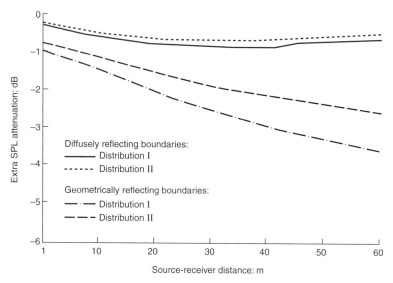

Figure 4.21. Extra SPL attenuation along the length caused by replacing distribution III with distributions I and II

II, and $0.791^3 = 0.49$ with distribution III. From Figure 4.21 it is also apparent that with geometrical boundaries the differences between the three distributions are greater than those with diffuse boundaries. This is probably because geometrical boundaries are more affected by the reflection pattern described above. It is noted that in the above calculation, if the absorption coefficient of the hard boundaries becomes higher, the difference between various distributions may become less. Moreover, if both façades are acoustically hard and geometrically reflective, the sound attenuation may be diminished by the multiple reflections between them.

4.2.6. Absorption from air and vegetation

At relatively high frequencies, air absorption may considerably reduce the sound level in urban streets. The absorption from vegetation may have a similar effect. For a typical street of $b = 20$ m and $a = 18$ m, the extra SPL attenuation caused by air absorption is shown in Figure 4.22, for both diffusely

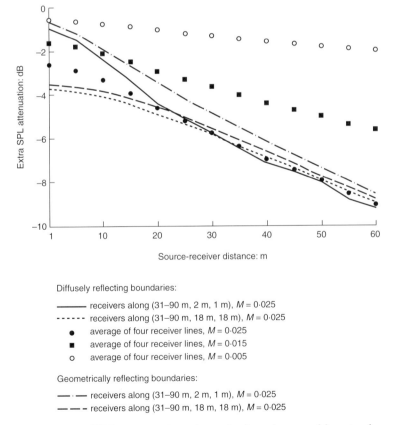

Diffusely reflecting boundaries:

———— receivers along (31–90 m, 2 m, 1 m), $M = 0.025$
- - - - - receivers along (31–90 m, 18 m, 18 m), $M = 0.025$
●   average of four receiver lines, $M = 0.025$
■   average of four receiver lines, $M = 0.015$
○   average of four receiver lines, $M = 0.005$

Geometrically reflecting boundaries:

—·— receivers along (31–90 m, 2 m, 1 m), $M = 0.025$
— —– receivers along (31–90 m, 18 m, 18 m), $M = 0.025$

Figure 4.22. Extra SPL attenuation along the length caused by air absorption

and geometrically reflecting boundaries. In the calculation the air absorption is $M = 0.025$ Np/m, the source is at (30 m, 6 m, 1 m) and the receivers are along two lines, namely (31–90 m, 2 m, 1 m) and (31–90 m, 18 m, 18 m). From Figure 4.22 it can be seen that the effect of air absorption is significant. It is note-worthy that with diffusely reflecting boundaries the extra SPL attenuation caused by air absorption is systematically greater than that with geometrically reflecting boundaries, although the difference is only within 1 dB. An important reason is that with diffusely reflecting boundaries the sound path is generally longer and, thus, air absorption is more effective.

To demonstrate the effect of M, Figure 4.22 also compares the extra attenua-tion with $M = 0.005, 0.015$ and 0.025 Np/m, where the boundaries are diffusely reflective and the SPL presented is based on the average of four receiver lines, namely (31–90 m, 2 m, 1 m), (31–90 m, 2 m, 18 m), (31–90 m, 18 m, 1 m) and (31–90 m, 18 m, 18 m). The three M values correspond approximately to the air absorption at 3000, 6000 and 8000 Hz, at a temperature of 20°C and relative humidity of 40–50% [4.7]. From Figure 4.22 it can be seen that with this range of M, the variation in sound attenuation is about 2–7 dB.

4.2.7. Multiple sources

The above calculations are all based on a single source. This is useful to gain a basic understanding of sound propagation in street canyons. The results are representative of certain types of urban noise, such as low-density traffic. They are also useful for considering noise propagation from a junction to a street. However, in many cases it is necessary to take multiple sources into account.

Consider a street canyon with evenly distributed point sources along the length. The street is 120 m long and the source spacing is σ. For convenience, assume that with a single source at any location, the SPL attenuation from the cross-section containing the source to a source-receiver distance of 60 m is D dB, and that this attenuation is linear along the length. Also assume that an acoustic treatment, such as increasing boundary absorption, can bring an extra attenuation of Q_0 dB in the cross-section with the source and an extra attenuation of Q_{60} dB at 60 m, and that this extra attenuation also increases linearly along the length. With these simplifications, the extra SPL attenuation caused by a treatment in the case of multiple sources can be readily calculated using MUL (see Section 2.6.4). Table 4.1 shows typical relationships between the extra attenuation with a single source and with multiple sources. Two typical receivers are considered in the case of multiple sources: $x = 60$ m, with no horizontal distance from a source; and $x = 60 + \sigma/2$, halfway between two sources. With multiple sources, as expected, sources near the receiver are more important than the others and, thus, the extra SPL attenuation caused by a given treatment is generally less than that with a single source. However, in Table 4.1 the extra attenuation with multiple sources is still significant. In

Table 4.1. Relationships between the extra SPL attenuation with a single source and with multiple sources

Extra attenuation with a single source (dB)			Extra attenuation with multiple sources (dB)			
D	Q_0	Q_{60}	$\sigma = 5\,\mathrm{m}$ $x = 60\,\mathrm{m}$	$\sigma = 15$ $x = 60$	$\sigma = 5$ $x = 60 + \sigma/2$	$\sigma = 15$ $x = 60 + \sigma/2$
10	2	2	2·0	2·0	2·0	2·0
10	2	5	2·9	2·9	2·9	2·9
10	2	10	4·2	4·0	4·2	4·3
10	−0·5	10	2·2	2·0	2·2	2·4
15	2	2	2·0	2·0	2·0	2·0
15	2	5	2·7	2·6	2·7	2·8
15	2	10	3·7	3·5	3·8	3·9
15	−0·5	10	1·6	1·3	1·7	1·9

other words, if an acoustic treatment is effective with a single source, it is also effective with multiple sources. The effectiveness with multiple sources increases with increasing D, Q_0 and Q_{60}.

Since a slight SPL increase may occur in the near field with some acoustic treatments, such as replacing geometrically reflecting boundaries with diffusely reflecting boundaries, a negative value of Q_0 is considered in Table 4.1. As expected, with a negative Q_0 the extra SPL attenuation with multiple sources becomes less.

4.2.8. Reverberation

A comparison of the RT30 and EDT between diffusely and geometrically reflecting boundaries is shown in Figure 4.23. The calculation is carried out with a street width of 20 m, three street heights of 6 m, 18 m and 30 m, a point source at (30 m, 6 m, 1 m) and receivers along (31–90 m, 2 m, 1 m). It is interesting to note that, in comparison with diffuse boundaries, the EDT with geometrical boundaries is shorter with a relatively low width/height ratio and is longer when this ratio is relatively high, whereas the RT30 is much longer with any aspect ratio. An important reason for the difference in RT30 is that, with geometrical boundaries, the image sources are well sepa-rated (see Figure 2.15) and, thus, the ratio of initial to later energy at a receiver is less than that with diffuse boundaries. Similar phenomenon occurs in long enclosures with a strongly absorbent ceiling, where the reverberation can be reduced by putting diffusers on smooth boundaries (see Chapter 5). From Figure 4.23 it can also be seen that, for both diffuse and geometrical bound-aries, the RT30 and EDT increase with the increase of source-receiver distance, which is again similar to the situation in long enclosures, as described in Chapters 2 and 3.

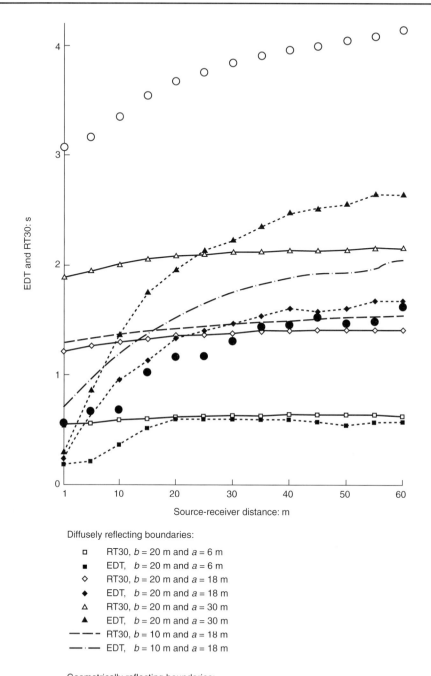

Diffusely reflecting boundaries:

- □ RT30, b = 20 m and a = 6 m
- ■ EDT, b = 20 m and a = 6 m
- ◇ RT30, b = 20 m and a = 18 m
- ◆ EDT, b = 20 m and a = 18 m
- △ RT30, b = 20 m and a = 30 m
- ▲ EDT, b = 20 m and a = 30 m
- — — - RT30, b = 10 m and a = 18 m
- — · — EDT, b = 10 m and a = 18 m

Geometrically reflecting boundaries:

- ○ RT30, b = 20 m and a = 6–30 m
- ● EDT, b = 20 m and a = 6–30 m

Figure 4.23. Comparison of reverberation time between diffusely and geometrically reflecting boundaries

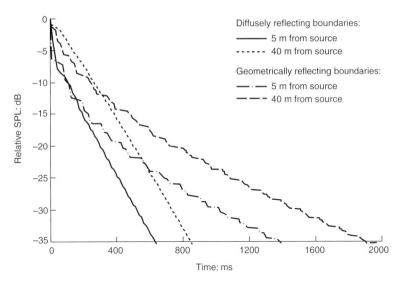

Figure 4.24. Comparison of decay curves between diffusely and geometrically reflecting boundaries

Corresponding to Figure 4.23, Figure 4.24 shows the decay curves at two typical receivers, 5 m and 40 m from the source, where the street height is 18 m. From the decay curves the above mentioned differences between diffuse and geometrical boundaries and the differences between near and far fields can be clearly seen.

The effect of street height on reverberation is similar to that on SPL. In Figure 4.23 it is seen that with geometrical boundaries, because S_z and R_z are less than 6 m, the reverberation times are constant with the three street heights, whereas with diffuse boundaries the RT30 and EDT become longer with the increase of street height.

Figure 4.23 also demonstrates the effect of street width on reverberation in the case of diffuse boundaries, as can be seen by comparing cross-sections 20 m by 18 m and 10 m by 18 m. As expected, with a given street height the RT30 and EDT are shorter with a greater street width because sky absorption is increased. In the case of geometrically reflecting boundaries, calculation shows that reverberation increases with increasing street width [4.8].

The distribution of RT30 and EDT in a cross-section can be roughly examined by calculating the ratio between two typical receivers in the cross-section. For a street with $b = 20$ m and $a = 18$ m, Figure 4.25 shows the ratios between receivers (31–90 m, 2 m, 1 m) and (31–90 m, 18 m, 18 m). The result suggests that for both diffusely and geometrically reflecting boundaries the RT30 is rather even throughout all the cross-sections along the length, whereas the EDT is only even in the cross-sections beyond a certain distance from the

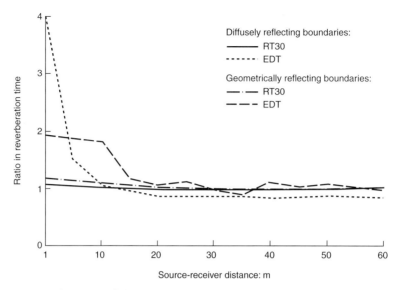

Figure 4.25. The ratio of the reverberation time at receivers (31–90 m, 2 m, 1 m) to receivers (31–90 m, 18 m, 18 m)

source. This distance appears to be slightly shorter with diffusely reflecting boundaries than with geometrically reflecting boundaries.

As expected, calculation shows that reverberation becomes shorter if air absorption is considered. It is noted that the reduction in reverberation caused by air absorption is more significant than that in SPL. As mentioned previously, this is because the SPL depends mainly on early reflections, whereas reverberation is dependent on multiple reflections, for which air absorption is more effective due to the longer sound path.

4.2.9. Discussion

The real world is generally somewhere between the sound fields under the two idealised boundary conditions compared above [4.9,4.10]. Consider a general condition, namely, when a sound ray is incident on a boundary, a fraction U_g of the incident power is reflected geometrically, a fraction U_d is reflected diffusely and the rest is absorbed. A street that consists of a mixture of boundaries, some geometrically reflective and some diffusely reflective, can be represented using $U_g = 0$ or $U_d = 0$ for each boundary. For a rectangular street canyon with such a general condition, it can be demonstrated that at a given source-receiver distance the average length of reflected sound path and the average order of reflection are between those resulting from diffuse and geometrical boundaries. Consequently, the true sound field should be between the two limits. As a calculation example, Davies has shown that, with $U_g = 0.6$ and $U_d = 0.3$, the attenuation curve along the length is between those with $U_g = 0.3$ and $U_d = 0.6$, and $U_g = 0.9$ and $U_d = 0$ [4.11]. Moreover, experimental evidence in Chapter 5 suggests that in long enclosures the sound attenuation along the length becomes systematically greater with an increasing amount of diffusers on boundaries or with better diffusers.

If most street boundaries are predominantly diffusely reflective, due to the effect of multiple reflections, the sound field in a street should be close to that resulting from purely diffusely reflecting boundaries. It has been demonstrated that if the façades in a street are diffusely reflective, there is no significant difference in the sound field of the street whether the ground is diffusely or geometrically reflective [4.12]. Moreover, there seems to be strong evidence that even untreated boundaries produce diffuse reflections [4.13].

4.3.   Interconnected streets

The models for calculating sound distribution in urban areas can be divided roughly into two groups, either 'microscopic' or 'macroscopic' in approach [4.14]. The former, often using simulation techniques, are used for accurately calculating the sound field in a street where buildings densely flank the roads, as described in the previous Sections. The latter, normally involving statistical methods and simplified algorithms, are for generally describing the sound distribution in a relatively large urban area [4.15]. An important link between the two kinds of approach is the behaviour of sound at street junctions and the sound propagation from one street to an intersecting street. This can lead to an improved understanding of noise control in a network of inter-connecting streets typical of urban areas. In this Section an urban element consisting of a major street and two side streets is considered using the radiosity model, in order to examine the effectiveness of various boundary treatments and urban design options on noise reduction [4.16,4.17].

The basic configuration used in the parametric study is illustrated in Figure 4.26. The size of the urban element is 120 m by 120 m. The element is divided into five areas, which are called streets N, S, W, E and M, respectively. In the calculation, each boundary is divided into 400–600 patches. This allows the program to calculate the form factors accurate to four decimal places. The diffraction over buildings is ignored because in the configurations considered, the energy transferring through street canyons is dominant. Most calculations are carried out with a single omnidirectional source but the situation with multiple sources is also considered. The sound attenuation along the length is based on the average of 5–10 receivers across the width. The source and receiver heights are 1 m.

4.3.1.   SPL and reverberation with a typical configuration

First a calculation is made with a typical configuration, where the street width and height are 20 m and the absorption coefficient of all the boundaries is 0·1.

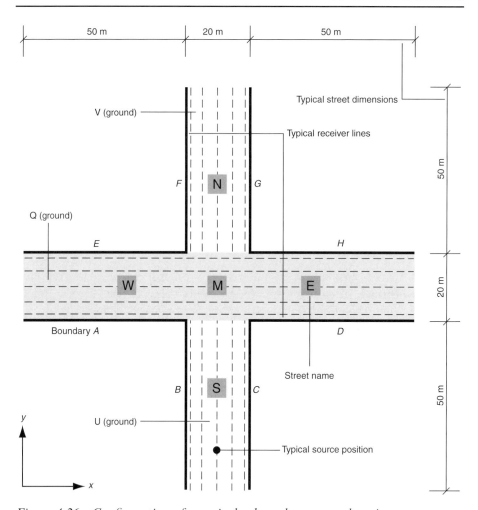

Figure 4.26. Configuration of a typical urban element — plan view

Figure 4.27(a)–(e) shows the SPL distribution with a point source at five positions in streets S and M: (60 m, 0 m, 1 m), (60 m, 15 m, 1 m), (60 m, 30 m, 1 m), (60 m, 45 m, 1 m) and (60 m, 60 m, 1 m). As expected, when the source is closer to the middle of the street junction, the average SPL in the streets becomes higher because less energy from the source can be reflected out of the streets. In street W or E, for example, the average SPL difference between source positions (60 m, 0 m, 1 m) and (60 m, 60 m, 1 m) is 18 dB, which is significant.

Based on the data in Figure 4.27, the SPL distribution with multiple sources can be obtained using MUL. Figure 4.28 shows such a distribution, where the sources are along $y=0$–120 m and with a spacing of 15 m. The result can also be approximately regarded as the Leq when a single source is moving from (60 m, 0 m, 1 m) to (60 m, 120 m, 1 m). Note, in this case the SPL scale

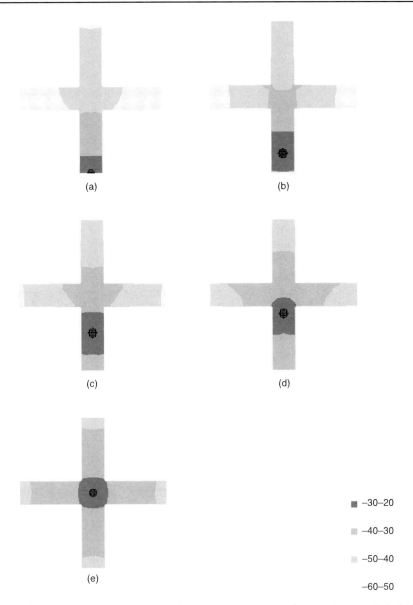

(a)

(b)

(c)

(d)

(e)

■ −30–20

▨ −40–30

▨ −50–40

−60–50

Figure 4.27. SPL distribution with a point source at five positions: (a) (60 m, 0 m, 1 m); (b) (60 m, 15 m, 1 m); (c) (60 m, 30 m, 1 m); (d) (60 m, 45 m, 1 m); (e) (60 m, 60 m, 1 m)

in Figure 4.28 should be approximately changed to −20 to −70 dB because the number of sources is reduced from 9 to 1. From Figure 4.28 it can be seen that the average SPL in street W or E is considerably lower than that in street S-M-N, at 9 dB on average. Also, the SPL attenuation along street W or E is significant, at about 15–17 dB from $x = 50$ m to 0 m or from $x = 70$ m

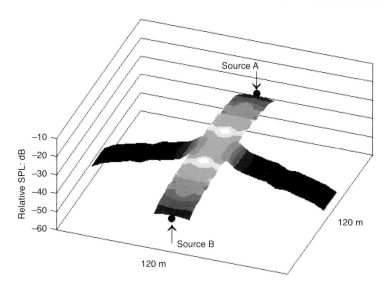

Figure 4.28. Three-dimensional SPL distribution with nine sources evenly distributed from (60 m, 0 m, 1 m) (point A) to (60 m, 120 m, 1 m) (point B), or with a single source moving from A to B

to 120 m. These results quantitatively demonstrate that if noise sources are along a major street like S-M-N, it is an effective way to reduce noise by arranging buildings in side streets like W or E.

To investigate the effect of side streets on the sound field in a major street, a comparison is made between three cases:

(*a*)   case GI, with the same configuration as above;
(*b*)   case GII, without streets W and E and the façades in street S-M-N are
        totally absorbent at $y = 50\text{--}70$ m; and
(*c*)   case GIII, street S-M-N only, with continuous façades.

A point source is at (60 m, 15 m, 1 m) in street S. In Figure 4.29 the three configurations are illustrated and the SPL attenuation along street S-M-N is compared. It is interesting to note that the difference in SPL attenuation between cases GI and GII is only about 0·5–1 dB and this is limited in the range 50–80 m. This means that the sound energy reflected from side streets W and E to the major street S-M-N is negligible. The effect of the boundaries at $y = 50\text{--}70$ m on street S-M-N can be seen by comparing cases GII and GIII. The difference is about 2 dB and this occurs in the range $y = 40\text{--}120$ m.

Figure 4.30 shows the distribution of RT30 and EDT with three source positions from the end of the major street to the middle of the street junction, namely (60 m, 0 m, 1 m), (60 m, 30 m, 1 m) and (60 m, 60 m, 1 m). It can be seen that, except in the near field, the RT30 is about 1–2 s and the EDT

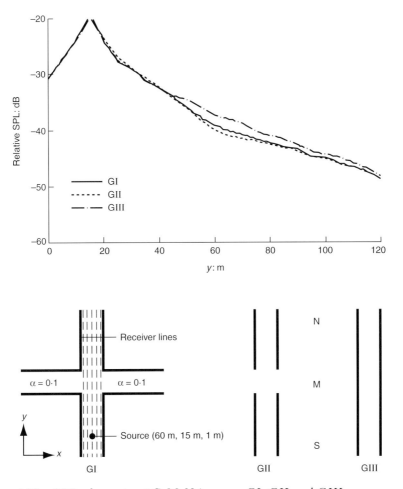

Figure 4.29. SPL along street S-M-N in cases GI, GII and GIII

varies from 0·2 s to 3 s. This suggests that the reverberation effect is significant
in the streets. Generally speaking, both the RT30 and EDT become longer with
increasing source-receiver distance. When the source moves from the end of the
major street to the middle of the street junction, the average RT30 in all the
streets increases slightly, at about 10%. In streets W and E the reverberation
is systematically longer than that in street S-M-N. This is particularly
significant for EDT, which can be seen in Figures 4.30(a′) and 4.30(b′). The
major reason for the long reverberation in streets W and E is the lack of
direct sound.

4.3.2. Street aspect ratio
With increasing street width/height ratio, more energy can be reflected out of
the street canyons and, thus, the overall SPL in the streets should become

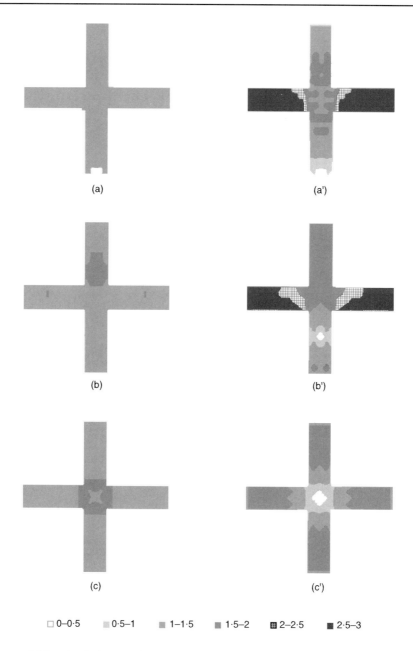

(a)   (a')

(b)   (b')

(c)   (c')

☐ 0–0·5   ▨ 0·5–1   ▨ 1–1·5   ▨ 1·5–2   ▦ 2–2·5   ▰ 2·5–3

Figure 4.30. RT30(s) and EDT(s) distribution with a point source at three positions — RT30: (a) (60 m, 0 m, 1 m); (b) (60 m, 30 m, 1 m); (c) (60 m, 60 m, 1 m) — EDT: (a′) (60 m, 0 m, 1 m); (b′) (60 m, 30 m, 1 m); (c′) (60 m, 60 m, 1 m)

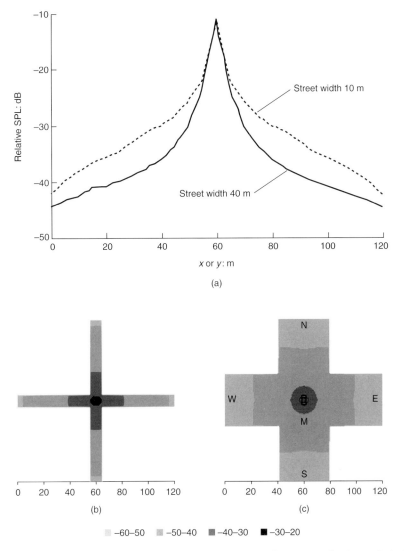

Figure 4.31. Comparison between street widths of 10 m and 40 m: (a) SPL attenuation along street S-M-N or W-M-E; (b) SPL distribution with street width of 10 m; (c) SPL distribution with street width of 40 m

lower. Figure 4.31(a) compares the SPL attenuation along the street centre (0–120 m, 60 m, 1 m) between two street widths, 10 m and 40 m. This corresponds to a change in street width/height ratio from 0·5 to 2. In the calculation the source is at (60 m, 60 m, 1 m), the street height is 20 m and the absorption coefficient of all the boundaries is 0·1. From Figure 4.31(a) it can be seen that with a street width of 40 m the attenuation is about 3–8 dB greater than that with a street width of 10 m, except in the very near field where the direct

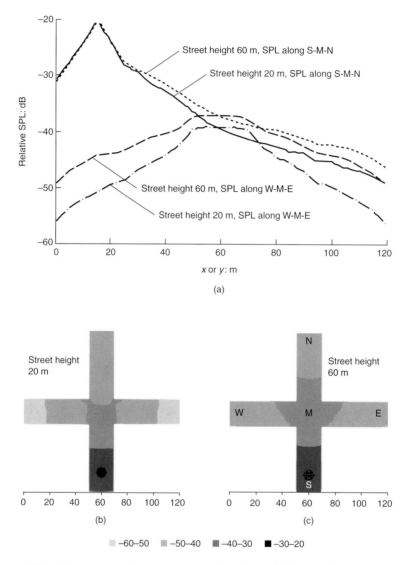

Figure 4.32. Comparison between street heights of 20 m and 60 m: (a) SPL attenuation along streets S-M-N and W-M-E; (b) SPL distribution with street height of 20 m; (c) SPL distribution with street height of 60 m

sound is dominant. The differences in SPL distribution between the two street widths are shown in Figures 4.31(b) and 4.31(c).

The sound fields with two street heights, 20 m and 60 m, are compared in Figure 4.32, where a point source is at (60 m, 15 m, 1 m), the boundary absorption coefficient is 0·1 and the street width is 20 m. The street width/ height ratios in the two cases are 1 and 0·33, respectively. The comparison is for the SPL attenuation along streets S-M-N and W-M-E, as well as for the

SPL distribution. From Figure 4.32 it can be seen that the SPL is increased systematically by the increased street height, typically at about 3–6 dB. The increase becomes greater with increasing source-receiver distance. In street W-M-E, the SPL increase is greater than that in street S-M-N. This is probably because in street S-M-N the direct sound plays an important role, whereas in street W-M-E the sound field is dominated by reflected energy.

Figure 4.33 compares two cases: case SI, where street S is 10 m wide and street N is 30 m wide; and case SII, where street S is 30 m wide and street N

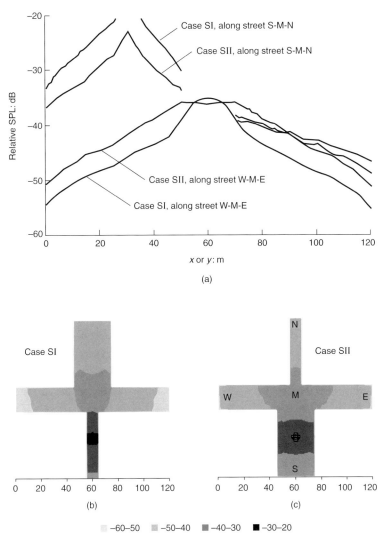

Figure 4.33. Effectiveness of changing street cross-section — comparison between cases SI and SII: (a) SPL attenuation along streets S-M-N and W-M-E; (b) SPL distribution in case SI; (c) SPL distribution in case SII

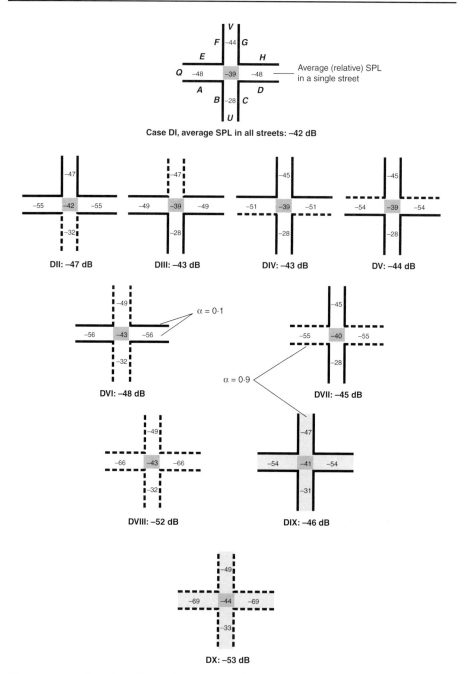

Figure 4.34. Relative SPL with various absorber arrangements. The number in each street is the average SPL in this street, and the bold numbers indicate the average SPL in the whole urban element. The continuous lines and white ground areas represent boundaries with $\alpha = 0.1$, and the broken lines and hatched ground areas represent boundaries with $\alpha = 0.9$

is 10 m wide. Again, the streets are 20 m high and the boundary absorption coefficient is 0·1. The source is at (60 m, 30 m, 1 m) in street S. In comparison with case SI, in case SII more sound energy can be reflected out of street S and, thus, the average SPL in this street is about 4 dB less. In the mean time, however, in case SII the SPL in streets W-M-E is generally increased, also by about 4 dB. Clearly this is because there is more energy from street S. In street N, the SPL difference between cases SI and SII is insignificant, generally within 1 dB.

4.3.3. Boundary absorption

To investigate the effect of boundary absorption, a calculation is made with absorbers at various positions in the urban element. Figure 4.34 shows ten typical absorber arrangements together with the average SPL in the whole element and in each street. In the calculation the absorption coefficient of the absorbers is 0·9, a point source is positioned at (60 m, 15 m, 1 m) and the street height is 20 m.

By comparing cases DII to DV, it can be seen that with a constant amount of absorption, 2×20 m $\times 50$ m, absorbers are more effective when they are arranged on boundaries B and C. In case DII the overall SPL is -47 dB, which is 5 dB lower than that in case DI and 3–4 dB lower than that in cases DIII to DV. An apparent reason for the low SPL in case DII is that on façades B and C the direct sound energy is stronger than that on other façades and, thus, the absorbers are more efficient. For the same reason, absorbers on boundaries E and H are more effective than those on A and D, especially for streets W and E, which can be seen by comparing cases DIV and DV.

By comparing cases DII, DVI, DVIII and DX, it can be seen that the overall SPL decreases with increasing boundary absorption. This is expected. However, it is noted that the SPL in street S is almost not affected. This means that the energy reflected back to street S from other streets is negligible. By comparing cases DIV, DV and DVII it is seen that the effects of the absorbers on boundaries A, D, E and H are also rather 'local'.

The effectiveness of ground absorption can be demonstrated by comparing cases DI and DIX, or cases DVIII and DX. As expected, the ground absorption is more effective when the façades are acoustically hard (compare cases DI and DIX) and becomes less efficient when the façades are absorbent (compare cases DVIII and DX).

With the same street layout as above, calculation is carried out by evenly increasing the absorption coefficient of all the boundaries from 0·01 to 0·99. Two source positions are considered, one at (60 m, 60 m, 1 m) and the other at (60 m, 15 m, 1 m). Figure 4.35 shows the average SPL of all the streets with increasing boundary absorption coefficient. It can be seen that when the

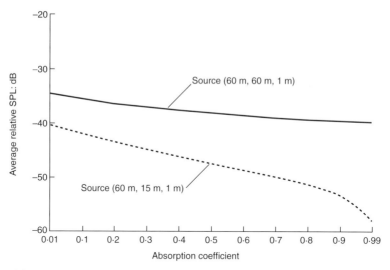

Figure 4.35. Average SPL in the urban element with various boundary absorption coefficients — two source positions are considered

source is at (60 m, 15 m, 1 m), the SPL decreases by 18 dB from $\alpha = 0\cdot01$ to $0\cdot99$, and by 11 dB from $\alpha = 0\cdot1$ to $0\cdot9$ (also compare cases DI and DX in Figure 4.34). When the source is at (60 m, 60 m, 1 m), the SPL decrease from $\alpha = 0\cdot01$ to $0\cdot99$, which is only about 5–6 dB. An important reason for the difference between the two source positions is that with the source at the middle of the street junction, (60 m, 60 m, 1 m), the direct sound plays a dominant role in a considerable area and, thus, the boundary absorption is relatively less efficient. Overall, the results in Figure 4.35 demonstrate the effectiveness of boundary absorption for reducing noise.

4.3.4. Staggered street

If a noise source is in street S, the SPL in street N may be reduced by staggering the two streets in x direction because this can diminish the direct sound, lengthen the reflection path and increase the number of reflections. Figure 4.36 compares the SPL in three cases:

(*a*)   case RI, with the same configuration as Figure 4.27(b);
(*b*)   case RII, street N is shifted to $x = 70$–90 m; and
(*c*)   case RIII, street S is shifted to $x = 30$–50 m and street N is at $x = 70$–90 m.

A point source is positioned in street S. It is in the middle of the street width and with $y = 15$ m. From Figure 4.36 it can be seen that by staggering streets S and N, a significant extra attenuation can be obtained in street N,

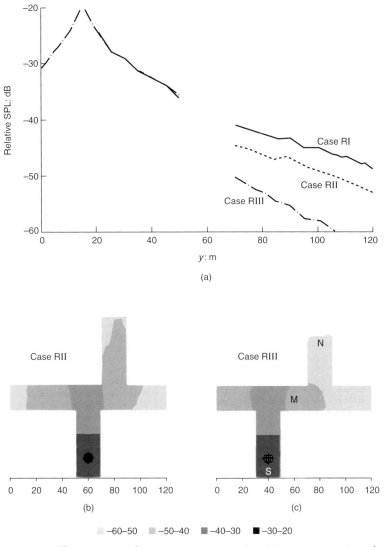

Figure 4.36. Effectiveness of staggering street S-M-N — comparison between case RI (same as Figure 4.27(b)), case RII and case RIII: (a) SPL attenuation along street S-M-N; (b) SPL distribution in case RII; (c) SPL distribution in case RIII

which is about 4–5 dB in case RII and 10–15 dB in case RIII. Conversely, in street S the difference in SPL between the three cases is unnoticeable. In street W-M-E, the SPL distribution pattern varies in the three cases but the average SPL is almost constant. Overall, the results in Figure 4.36 demonstrate the effectiveness of strategically planning the streets.

4.4. Summary

The parametric study demonstrates that considerable noise reduction can be obtained by strategic architectural design and urban planning. To provide a concise design guidance, the calculation results are summarised below. Note the conclusions regarding the urban element consisting of a major street and two side streets are based on diffuse boundaries.

4.4.1. Street configuration

(a) When changing the width/height ratio of a single street or an urban element consisting of a major street and two side streets, say, from 0·3 to 3, the variation in average SPL is typically 3–8 dB.

(b) Similarly, considerable extra attenuation along the length can be obtained by widening a street. Note that in this case the extra SPL attenuation is not monotonic along the length for both diffusely and geometrically reflecting boundaries.

(c) With multiple sources or a moving source along a major street, the average SPL in a side street is typically 9 dB lower than that in the major street. In the mean time, the SPL attenuation along the side streets is significant, at about 15–17 dB.

(d) Conversely, despite the significant changes in the boundary condition in the side streets, the SPL variation in the major street is only about 1 dB. This suggests that with noise sources along a major street, the energy reflected from side streets to the major street is negligible.

(e) In the urban element consisting of a major street and two side streets, if noise sources are in one side street, considerable extra sound attenuation, typically 5–15 dB, can be obtained in the other side street if the two streets are staggered.

(f) A gap between buildings can provide about 2–3 dB extra sound attenuation along the street and the effect is more significant in the vicinity of the gap.

(g) For both diffuse and geometrical boundaries the sound distribution in a cross-section is generally even unless the cross-section is very close to the source.

4.4.2. Absorption

(a) An absorption treatment in a single street, such as increasing the boundary absorption coefficient from 0·1 to 0·5, or taking one side of buildings away, or treating the ground as totally absorbent, can typically bring an extra attenuation of 3–4 dB. With very high absorption coefficient an extra attenuation of 7–13 dB can be achieved.

(b) In the urban element consisting of a major street and two side streets, the effectiveness of boundary absorption on reducing noise typically varies between 6–18 dB, depending on the source position.

(c)   With a given amount of absorption, for both diffusely and geometrically reflecting boundaries the sound attenuation along a single street is the highest if the absorbers are arranged on one façade and the lowest if they are evenly distributed on all boundaries. This result is especially useful for the design of a street canyon formed by noise barriers.

(d)   For the urban element consisting of a major street and two side streets, with a given amount of absorption, if the absorbers are arranged near the source and on the boundaries with strong direct sound energy, the average SPL in the streets can be 4 dB lower than that with other absorber arrangements.

(e)   With diffusely reflecting boundaries the extra attenuation caused by increasing boundary absorption is almost constant along the street length, whereas with geometrically reflecting boundaries the extra attenuation increases with the increase of source-receiver distance.

(f)   Air absorption is effective for increasing sound attenuation along the length, typically by 3–6 dB. Absorption from vegetation should have a similar effect.

4.4.3. Reflection characteristics of boundaries

There are considerable differences between the sound fields resulting from diffusely and geometrically reflecting boundaries. By replacing diffuse boundaries with geometrical boundaries:

(a)   the sound attenuation along the length becomes considerably less, typically by 4–8 dB with a source-receiver distance of 60 m;

(b)   the RT30 is significantly longer, typically by 100–200%; and

(c)   the extra SPL attenuation caused by air or vegetation absorption is reduced.

With moving traffic, about 2–4 dB extra attenuation can be obtained by using diffusely instead of geometrically reflecting boundaries.

The above results suggest that, from the viewpoint of urban noise reduction, it is better to design the building façades and the ground of a street canyon as diffusely reflective rather than acoustically smooth. Although it might be unrealistic to design all the boundaries as purely diffusely reflective, some diffuse patches on a boundary, or boundaries with a high diffusion coefficient, are helpful in making the sound field closer to that resulting from diffuse boundaries, especially when multiple reflections are considered. In a similar way to diffuse boundaries, street furniture, such as trees, lamp posts, fences, barriers, benches, telephone boxes and bus shelters, can act as diffusers and, thus, be effective in reducing noise. With the same principle as for a single street canyon, diffuse boundaries and street furniture are also useful for reducing the 'background' noise of a city, which is produced by the general distribution of sources throughout the city.

4.4.4.   Street reverberation

(a)   With a boundary absorption coefficient of 0·1, the reverberation time is typically 0·7–2 s in a single street and 1–3 s in the urban element consisting of a major street and two side streets. This indicates the importance of considering reverberation in urban streets.

(b)   With diffusely reflecting boundaries the reverberation in a street canyon becomes longer with a greater street height and becomes shorter with a greater street width. For both diffusely and geometrically reflecting boundaries the reverberation increases systematically with increasing distance from the source.

(c)   With a sound source in the major street of the urban element, the reverberation in the side streets is systematically longer than that in the major street. This is particularly significant for the EDT.

(d)   With both kinds of boundary the RT30 is rather even throughout all the cross-sections along the length, whereas the EDT is only even in the cross-sections beyond a certain distance from the source.

(e)   The change in reverberation caused by air absorption is more significant than that in SPL.

It is noted that the effectiveness of the above mentioned architectural changes and urban design options may be diminished if a receiver is very close to high-density traffic. Nevertheless, if a barrier is inserted to reduce the direct sound, it is still important to consider these design strategies. With multiple sources the extra SPL attenuation caused by a given acoustic treatment is generally less than that with a single source but is still significant.

4.5.   References

4.1   KANG J. Experiments on the subjective assessment of noise reduction by absorption treatments. *Chinese Noise and Vibration Control*, 1988, No. 5, 20–28 (in Chinese).

4.2   SCHRÖDER E. Nachhall in geschlossenen bebauten Straßen. *Lärmbekämpfung*, 1973, **17**, 11–18.

4.3   STEENACKERS P., MYNCKE H. and COPS A. Reverberation in town streets. *Acustica*, 1978, **40**, 115–119.

4.4   KO N. W. M. and TANG C. P. Reverberation time in a high-rise city. *Journal of Sound and Vibration*, 1978, **56**, 459–461.

4.5   KANG, J. Sound propagation in street canyons: comparison between diffusely and geometrically reflecting boundaries. *Journal of the Acoustical Society of America*, 2000, **107**, 1394–1404.

4.6   KANG J. Modelling the acoustic environment in city streets. *Proceedings of the PLEA 2000*, Cambridge, England, 512–517.

4.7   AMERICAN NATIONAL STANDARDS INSTITUTE (ANSI). *Method for the calculation of the absorption of sound by the atmosphere*. ANSI S1.26. ANSI, 1995 (Revised 1999).

4.8   KANG J. Sound propagation in urban streets: Comparison between UK and Hong Kong. *Proceedings of the 8th International Congress on Sound and Vibration*, Hong Kong, 2001, 1241–1248.

4.9   WU S. and KITTINGER E. On the relevance of sound scattering to the prediction of traffic noise in urban streets. *Acustica/Acta Acustica*, 1995, **81**, 36–42.

4.10 KUTTRUFF H. Stationäre Schallausbreitung in Langräumen. *Acustica*, 1989, **69**, 53–62.

4.11 DAVIES H. G. Multiple-reflection diffuse-scattering model for noise propagation in streets. *Journal of the Acoustical Society of America*, 1978, **64**, 517–521.

4.12 KANG J. Numerical modelling of the sound field in urban streets with diffusely reflecting boundaries. *Journal of Sound and Vibration* (submitted for publication).

4.13 HODGSON M. Evidence of diffuse surface reflections in rooms. *Journal of the Acoustical Society of America*, 1991, **89**, 765–771.

4.14 KUTTRUFF H. Zur Berechnung von Pegelmittelwerten und Schwankungsgrößen bei Straßenlärm. *Acustica*, 1975, **32**, 57–69.

4.15 BRADLEY J. S. A study of traffic noise attenuation around buildings. *Acustica*, 1977, **38**, 247–252.

4.16 KANG J. Sound propagation in interconnected urban streets: a parametric study. *Environment and Planning B: Planning and Design*, 2001, **28**, 281–294.

4.17 KANG J. Sound field in urban streets with diffusely reflecting boundaries. *Proceedings of the Institute of Acoustics (IOA) (UK)*, Liverpool, 2000, **22**, No. 2, 163–170.

5. Case study: design guidance based on scale modelling

In comparison with theoretical/computer models, physical scale models are more suitable for investigating relatively complicated configurations. The usefulness of scale modelling of long enclosures has been demonstrated by a number of practical applications [5.1–5.4]. This Chapter describes a series of measurements in two 1:16 scale models of underground stations, one rectangular and the other of a circular cross-sectional shape. The main objective is to demonstrate the effectiveness of strategic architectural acoustic treatments, including absorbers, diffusers, reflectors, obstructions and their combinations. Particular attention is given to the speech intelligibility of multiple loudspeaker PA systems, measured by the STI — since the fire disaster at London King's Cross underground station on 9 December 1987, many underground companies are actively investigating the improvement of their station PA speech intelligibility and, at the same time, the reduction of noise. Note, in this Chapter dimensions and frequencies relate to full scale, except where indicated.

5.1. A rectangular long enclosure

5.1.1. The model and modelling technique

5.1.1.1. Scale model

The scale model was designed to generally simulate the new stations of the Hong Kong MTR, in particular those with a central platform [5.4–5.6]. Since in these stations there are screen doors between the platform and tracks, the model was basically a rectangular box, as illustrated in Figure 5.1. For ease of construction, the model was 4880 mm long (78·08 m full scale), representing approximately a third to a half of the MTR stations' length. Nevertheless, this length should be sufficient for investigating the effectiveness of architectural acoustic treatments, since for a multiple loudspeaker PA system the speech intelligibility at a receiver depends mainly on the effect of the loudspeakers within this range. The inner cross-section of the model

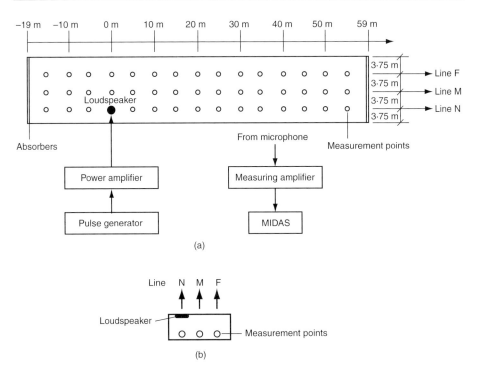

Figure 5.1. 1:16 scale model and the acoustic measurement system: (a) plan; (b) cross-section. The inner length, width and height of the model are 4880 mm (78·08 m full scale), 937·5 mm (15 m full scale) and 300 mm (4·8 m full scale), respectively

was 300 mm (4·8 m full scale) high and 937·5 mm (15 m full scale) wide, which was in correspondence with that of the MTR stations. For practicability, when altering ceiling arrangements the model was inverted for all the measurements.

Acoustically hard boundaries were modelled with well-varnished timber. To measure its absorption coefficient, a 1:16 model reverberation room was built with this model material. The measured absorption coefficient, as shown in Figure 5.2, was around 0·05 over the model frequency range. This corresponded to actual acoustically hard surfaces.

The design of the model was based on the experience gained from the scale modelling of London Euston Square underground station, a typical cut and cover station of rectangular cross-sectional shape [5.1–5.3]. For the SPL attenuation along the length, the agreement between full scale and model tests in the London Euston Square station was very good and generally within 2 dB accuracy at middle frequencies. For reverberation, the model RTs were very close to their full-size equivalents, generally within 10% accuracy at middle frequencies. For the STI, the differences between measured values in the scale model and at full scale were within ±0·05.

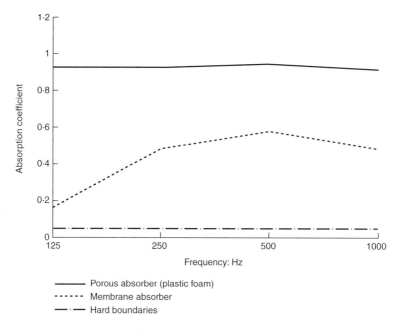

Figure 5.2. Absorption coefficients of the porous absorber (plastic foam), the membrane absorber and the hard boundaries measured in a model reverberation room

5.1.1.2. End walls

The model end walls were highly absorbent during all the measurements except those for investigating the effect of end walls. This was because the model only simulated part of the MTR stations and the model end walls were actually imaginary. If the model end walls were treated as reflective, either geometrically or diffusely, the sound energy in the model could be significantly higher than that in the actual stations. However, it was noted that by using absorbent model end walls, the scattering effect of boundaries beyond those modelled would be ignored. Fortunately, the effect of such boundaries has been demonstrated to be noticeable only within a limited range, namely in the area near the model end walls, and the actual end walls should be strongly absorbent for decreasing reverberation and increasing speech intelligibility, as analysed in Chapter 3.

The model absorber on the end walls was simulated by a 10 mm thick (16 cm full scale) plastic foam. Its absorption coefficient measured in the model reverberation room was around 0·9 over the model frequency range, as shown in Figure 5.2.

5.1.1.3. Model diffusers

A ribbed structural element, which can often be found in underground tunnels and which acts as a diffuser, was simulated. The model ribbed diffuser was a hard plastic grid with a spacing of 20 mm (32 cm full scale) between the grid lines. The thickness and width of the grid lines were 5 mm (8 cm full scale). For the convenience of arrangement, the model ribbed diffusers were in a size of 140 mm by 140 mm (2·24 m by 2·24 m full scale). Calculation showed that the effective scattering range of this ribbed diffuser was approximately above 500 Hz [5.7].

5.1.1.4. Model absorbers

Two typical absorbers, a porous absorber and a resonant absorber, were tested. The porous absorber was the same as that used on the model end walls. For the convenience of strategic arrangements, the model absorbers were in a size of 150 mm by 300 mm (2·4 m by 4·8 m full scale). There were two reasons for testing the resonant absorber. First, resonant absorbers might be less effective in long enclosures owing to the large angle of incidence in comparison with a diffuse field, especially at long distances. Second, in some cases non-fibrous materials must be used due to strict health requirements [5.8–5.11].

As a typical resonant absorber, a membrane structure was simulated. This was an airtight plastic film with an airspace from a hard boundary. The film was 0·01 mm thick and its surface density was 0·003 kg/m². In order to increase the acoustic resistance, the film was firmly stuck on plastic frames. The length, width and height of the frames were 3000 mm (48 m full scale), 100 mm (1·6 m full scale) and 4 mm (6·4 cm full scale), respectively. Due to natural droop, the airspace between the film and hard boundary was 2–4 mm (3·2–6·4 cm full scale). Measurements in the model reverberation room showed that the resonant frequency of this membrane absorber was approximately 8000 Hz (500 Hz full scale) and the corresponding absorption coefficient was around 0·5, as shown in Figure 5.2. This resonance range appeared to be in correspondence with the calculation by the following conventional formula

$$f_0 = \frac{600}{\sqrt{D_m L_m}} \tag{5.1}$$

where f_0 is the resonant frequency, D_m (kg/m²) is the surface density of the membrane and L_m (cm) is the depth of the airspace. Equation (5.1) has also been demonstrated to be fairly accurate for other kinds of membrane [5.11].

5.1.1.5. Model obstructions

Two common obstructions in long enclosures were investigated. One was barriers, which are often used to construct certain small spaces, such as rest areas in underground stations. The other was blocks, such as attendant cubicles in underground stations. The model barriers were simulated by well-varnished timber with a width of 234·5 mm (3·75 m full scale) and a thickness of 5 mm (8 cm full scale). The barrier height was the same as that of the model, namely 300 mm (4·8 m full scale). The model blocks were constructed out of the barriers.

5.1.1.6.   Measurement system

The measurements were carried out using MIDAS, a frequently adopted software for all-scale room acoustics measurements [5.12,5.13]. With MIDAS a variety of acoustic indices can be measured and the excessive air absorption at high frequencies can be numerically compensated. The measurement system is illustrated in Figure 5.1.

The source was a single tweeter (KEF Type T27), which simulated a PA loudspeaker [5.2]. The test signal was an impulse with a repetition time of 2 s and a duration of 0·05 m/s. Since the model was rather wide, it was considered that two lines of ceiling loudspeakers were appropriate [5.6]. To simulate this arrangement, the source was positioned at a distance of 234·5 mm (3·75 m full scale) from one side wall and 703 mm (11·25 m full scale) from the other, as shown in Figure 5.1. Between the source and two end walls, the distances were 1220 mm (19·52 m full scale) and 3660 mm (58·56 m full scale). With this source position, since the distance from the source to either end wall was relatively large, the possible effect caused by treating the model's end walls as totally absorbent (as analysed previously) should not be significant within a considerable range of source-receiver distance, say, 0–50 m.

The receiver was an omnidirectional electret microphone, Knowles CA1759, which was positioned at a height of 94 mm (1·5 m full scale). Corresponding to the loudspeaker arrangement above, measurement points were along three lines, namely, lines N, M and F in Figure 5.1. On each line there were 15 measurement points with a spacing of 5 m. At these points, measurements were made in five octaves corresponding to the 125 Hz to 2000 Hz octaves in full scale.

5.1.1.7.   Calculation with multiple loudspeakers

Although the measurements were carried out using a single loudspeaker, the acoustic indices with multiple loudspeakers can be calculated by MUL (see Section 2.6.4). According to the measurement arrangement, the calculation was based on two lines of ceiling loudspeakers, as illustrated in Figure 5.3. On each line, 21 loudspeakers were considered with a spacing of 5 m. Correspondingly, in equations (2.113) to (2.115) $L(t)_{z_j,\eta_j}$ was used. Two typical receivers were considered (see Figure 5.3). One was point A, which had no horizontal distance from the central loudspeaker on line N. The other was point B, which was between two central loudspeakers. The receiver height was 1·5 m, which was in correspondence with the microphone height in the measurement. At these two points, the maximum source-receiver distance along the length was 50 m. With this range, as analysed previously, the assumption of MUL was reasonable. For the convenience of analysis, the STI

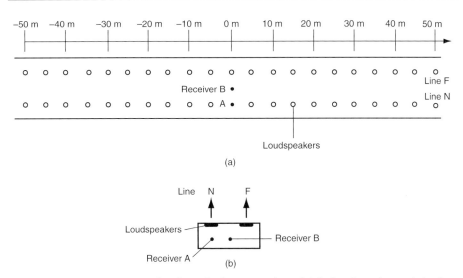

Figure 5.3. Arrangement for the calculation with multiple loudspeakers: (a) plan; (b) cross-section

calculation was made with an S/N ratio of 30 dB(A). In other words, the effect of background noise was not considered.

From the viewpoint of improving the speech intelligibility of a multiple loudspeaker PA system, architectural acoustic treatments should be designed to reduce the 'disturbance' between loudspeakers by increasing the sound attenuation along the length and by decreasing the reverberation from farther sources, as discussed in Section 3.4. Therefore, the following analysis focuses on the SPL attenuation and the EDT of the single loudspeaker and, correspondingly, the STI from multiple loudspeakers. Since the calculated STI from multiple loudspeakers is very close at points A and B, only the calculations at point A are presented. The results below are the arithmetic average of 500 Hz and 1000 Hz.

5.1.2. Empty condition

The variation in reverberation time and SPL along the length could be diminished by highly reflective end walls (see Section 3.2.5), especially when the length/cross-section ratio is relatively small. To verify this, measurements were first carried out with no absorbent treatment to any of the boundaries. Figures 5.4 and 5.5 show the EDT and SPL along line N under this empty condition.

It can be seen in Figures 5.4 and 5.5 that, except for the near field, the EDT variation and SPL attenuation along the length are not large. Moreover, the RT30/EDT ratio is generally close to 1, as shown in Figure 5.6, which means that the decay curves are rather linear. Furthermore, the RT calculated by the

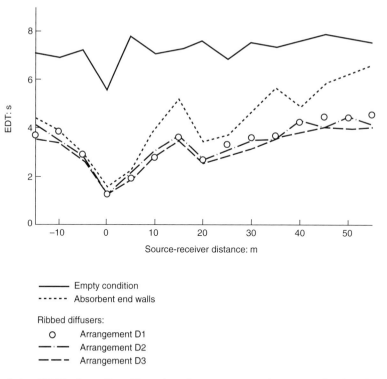

Figure 5.4. EDT along line N under the empty condition, with absorbent end walls, and with ribbed diffusers. Average of 500 Hz and 1000 Hz. The arrangements of diffusers are given in Figure 5.9, where the end walls are absorbent

Eyring formula is 6 s, which is close to the measured values. Overall, it appears that under the empty condition the sound field in the model is close to diffuse.

5.1.3. Absorbent end walls

In contrast with the above, the sound field is far from diffuse when the end walls are strongly absorbent. The comparison of the EDT along line N between reflective and absorbent end walls is shown in Figure 5.4. It can be seen that with absorbent end walls the EDT is much shorter. Clearly, this is due to the absorption in later reflections by the absorbent end walls. Moreover, with absorbent end walls the EDT increases significantly along the length. The peaks and troughs on the EDT curve, such as at 20 m and 40 m, are likely to be caused by the strong directionality of the source. Furthermore, with absorbent end walls, the RT30/EDT ratio decreases along the length, as shown in Figure 5.6. These results correspond to the theoretical analysis in Section 3.2.

The effect of absorbent end walls on the SPL is also significant, as shown in Figure 5.5. In comparison with reflective end walls, with absorbent end walls the SPL decreases systematically with increasing source-receiver distance.

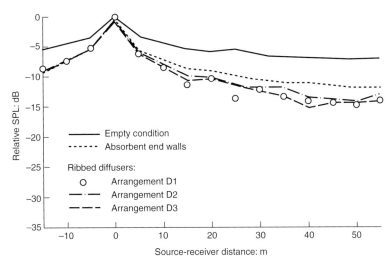

Figure 5.5. Sound attenuation along line N under the empty condition, with absorbent end walls, and with ribbed diffusers. Average of 500 Hz and 1000 Hz. The arrangements of diffusers are given in Figure 5.9, where the end walls are absorbent

This decrease is again caused by the absorption in multiple reflections by the absorbent end walls. However, it is noted that, since the SPL mainly depends on early reflections, the effect of absorbent end walls on the SPL is relatively less than that on the EDT.

To investigate the differences between lines N, M and F, the EDT and SPL along the three lines in the case of absorbent end walls are compared in

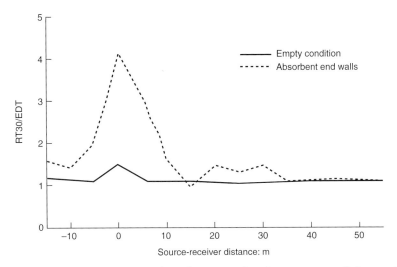

Figure 5.6. RT30/EDT ratios along line N under the empty condition and with absorbent end walls — average of 500 Hz and 1000 Hz

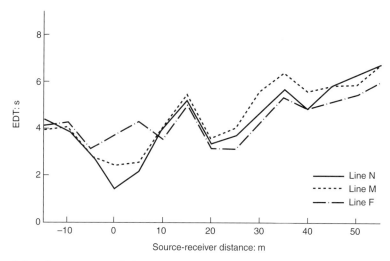

Figure 5.7. Comparison of the EDT along line N, line M and line F with absorbent end walls — average of 500 Hz and 1000 Hz

Figures 5.7 and 5.8. As expected, in a relatively near field, say, within 10 m, with a larger source-line distance the EDT is longer and the SPL is lower. Beyond this distance the differences between the three lines are much less. Similar variations between the three lines were observed in most of the other measurements. For this reason, the following presentations are mostly for line N only.

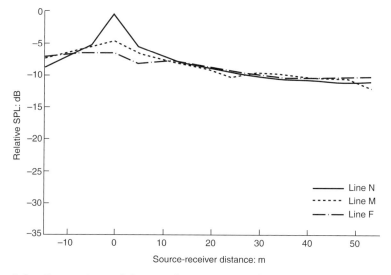

Figure 5.8. Comparison of the sound attenuation along line N, line M and line F with absorbent end walls — average of 500 Hz and 1000 Hz

From the above analysis it is clear that the absorbent treatment on end walls is very effective for improving speech intelligibility. This is more significant when the length/cross-section ratio is relatively small.

5.1.4. Ribbed diffusers

The effectiveness of the ribbed diffusers was investigated under both low and high absorption conditions. Under a low absorption condition, namely with no absorbers in the model except for the end walls, three diffuser arrangements between 0 m and 40 m were tested, as shown in Figure 5.9. These arrangements were: D1, 60 diffusers on the ceiling; D2, 44 diffusers on the ceiling; and D3, 44 diffusers on the ceiling and 8 diffusers on each side wall. Under a high absorption condition, namely with absorbers on both the end walls and two thirds of the ceiling, four diffuser arrangements were investigated, as shown in Figure 5.10. These arrangements were: D4, 60 diffusers from 0 m to 30 m, both on the ceiling and side walls; D5, 24 diffusers from 0 m to 30 m on side walls; D6, 60 diffusers along the ceiling; and D7, 60 diffusers along side walls.

Under the low absorption condition, the effectiveness of the ribbed diffusers on the EDT and SPL attenuation along line N is shown in Figures 5.4 and 5.5. From Figure 5.4 it can be seen that with diffusers the EDT is systematically lower, especially beyond a certain source-receiver distance, say, 10 m. A possible reason for this is that the sound energy, especially in later reflections, has more

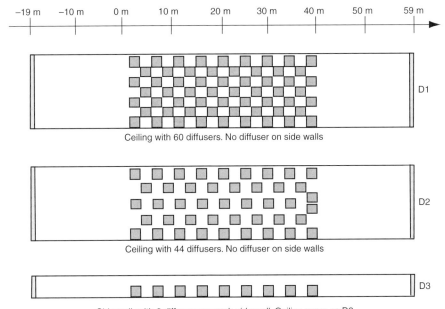

Figure 5.9. Experimental arrangements of the ribbed diffusers under the low absorption condition, namely, with no absorbers except for the end walls

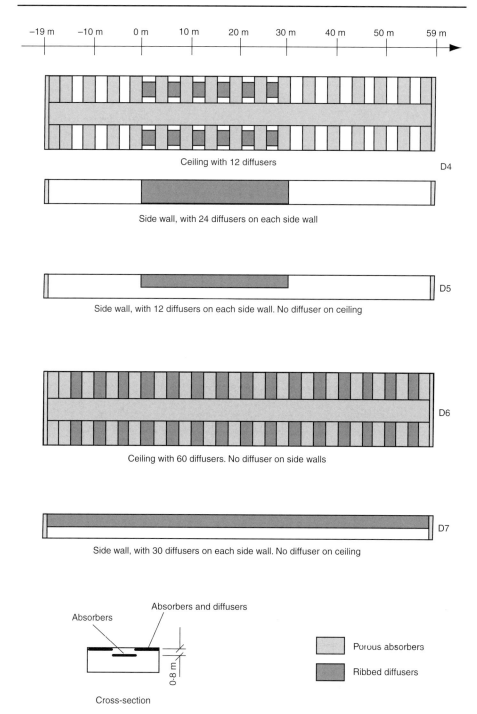

Figure 5.10. Experimental arrangements of the ribbed diffusers under the high absorption condition, namely, with absorbers on both the end walls and two thirds of the ceiling

opportunities to be absorbed. From Figure 5.5 it can be seen that, in correspondence with the theoretical analysis in Sections 2.1.1 and 3.1.4, with diffusers the sound attenuation along the length is systematically greater. Between $-15\,m$ and $0\,m$, the SPL is slightly higher than, or close to, that without diffusers. This means that diffusers can transfer sound energy to this area. Overall, with arrangements D1 to D3 the STIs from multiple loudspeakers are all greater than 0·30, which is significantly higher than that without diffusers, i.e. 0·20.

The effectiveness of the diffusers could be improved by increasing the number. In comparison with arrangement D2, with arrangement D3, namely 16 diffusers more on side walls, the sound attenuation along the length is greater and the EDT is shorter.

Correspondingly, the STI from multiple loudspeakers becomes 0·32 from 0·30. However, the STI is not always higher with more diffusers. As compared with arrangement D2, with arrangement D1, namely 16 diffusers more on the ceiling, the EDT is slightly longer beyond about 20 m. Consequently, the STI from multiple loudspeakers resulting from arrangement D1 is 0·31, which is even lower than that of arrangement D2. A possible reason for this increase in EDT is that the additional diffusers cause more energy reduction in early reflections than in later reflections.

Under the high absorption condition, the effectiveness of the ribbed diffusers is less than that under the low absorption condition but is still noticeable. Figures 5.11 and 5.12 show the EDT and SPL along line N resulting from arrangement D4. It can be seen that with diffusers the EDT is generally shorter and the sound attenuation along the length is systematically greater. Correspondingly,

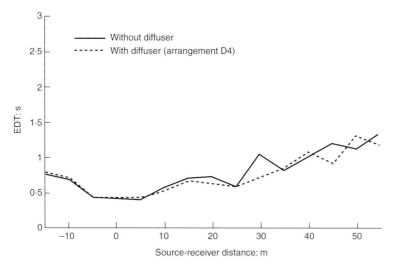

Figure 5.11. Effectiveness of the ribbed diffusers for the EDT along line N under the high absorption condition. Average of 500 Hz and 1000 Hz. Arrangements given in Figure 5.10

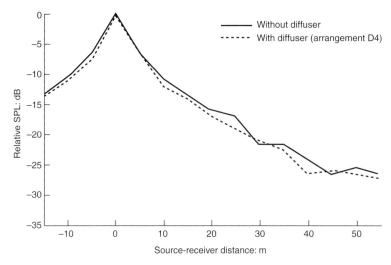

Figure 5.12. Effectiveness of the ribbed diffusers for the sound attenuation along line N under the high absorption condition. Average of 500 Hz and 1000 Hz. Arrangements given in Figure 5.10

the STI from multiple loudspeakers is 0·66, which is higher than that without diffusers, 0·64. This effectiveness, however, is considerably less than that under the low absorption condition. An important reason is that with highly absorbent boundaries the number of reflections becomes fewer and, thus, the efficiency of the diffusers is diminished. In other words, the direct sound plays a relatively important part. Similar results have also been obtained in another long enclosure, as described in Section 5.2. In Kuttruff's calculation [5.14] it was also shown that when the absorption coefficient was increased the difference in SPL between diffusely and geometrically reflecting boundaries was reduced.

With arrangements D5, D6 and D7, the STIs from multiple loudspeakers are 0·65, 0·65 and 0·64, respectively. By comparing the STIs of arrangements D4 to D7, it appears that the effectiveness of diffusers is diminished by two ways, one by fewer in number, e.g. from arrangement D4 to D5, and the other by lower density, e.g. from arrangement D4 to D6 or D7.

In summary, the above results demonstrate that strategically arranged ribbed diffusers are effective for increasing the sound attenuation along the length, for decreasing the reverberation from farther loudspeakers and, consequently, for improving the speech intelligibility of multiple loudspeaker PA systems. This effectiveness could be greater with a better diffuser, such as a Schroeder diffuser, as described in Section 5.2.1.

5.1.5. Porous absorbers

The investigation of the porous absorber focused on two aspects, namely, the amount and distribution of absorption. For the former, two absorber quantities, 64 and 96 absorbers, were tested. These corresponded to densities of 66% and 100% of the ceiling area. For the latter, six absorber arrangements, four with 64 absorbers and two with 96 absorbers, were compared. For the convenience of analysis, the six arrangements, as illustrated in Figure 5.13, are called arrangements A1 to A6. In Figures 5.14 and 5.15, the EDT and SPL along line N resulting from these arrangements are shown.

Firstly, arrangements A1 to A4, i.e. the arrangements with 64 absorbers, are compared. Between arrangements A1 and A2, it appears that there is no systematic difference for both the EDT and SPL. The calculated STIs from multiple loudspeakers are 0·63 for arrangement A1 and 0·64 for arrangement A2. With arrangement A3, generally speaking, the EDT is shorter than, and the sound attenuation along the length is greater than, those with arrangements A1 and A2. Correspondingly, the STI from multiple loudspeakers resulting from arrangement A3 is 0·67. A possible reason for this improvement is that with arrangement A3 some absorbers are suspended and, thus, more efficient. This efficiency could even be higher with arrangement A4, with which all the absorbers are vertical. From Figure 5.15 it can be seen that with arrangement A4 the sound attenuation along the length is much greater than that with arrangements A1 to A3. For example, at 50 m the SPL difference between arrangements A4 and A1 to A3 is 7–9 dB. Unfortunately, from Figure 5.14 it is noted that, with arrangement A4, the EDT is significantly longer than that with arrangements A1 to A3. As a result, the STI from multiple loudspeakers resulting from arrangement A4 is 0·59, which is even lower than that with arrangements A1 to A3.

The above increase in EDT resulting from arrangement A4 may be caused by two reasons. One is that the vertical absorbers may cause more energy reduction in early reflections than in later reflections. Second, due to the lack of absorption around the loudspeaker, the reflections, which are nearly vertical to the floor, may cause more later energy, particularly in a relatively near field. To investigate this effect, an absorber was put behind the loudspeaker in the model, as shown in Figure 5.13. It is interesting to note that in this case the EDT is systematically reduced, particularly within about 20 m (see Figure 5.14). In contrast, the effect of the added absorber on the sound attenuation along the length is unnoticeable (see Figure 5.15). This is because the SPL depends mainly on early reflections whereas the added absorber is more effective on later reflections.

The above example demonstrates the importance of acoustic design in the near field. Moreover, in some measurements, standing waves between ceiling and floor have been observed at 0 m. This may seriously decrease the speech intelligibility at this point. To avoid this effect, the ceiling around loudspeakers should be designed carefully. This may involve using absorbent or diffuse surfaces, a ceiling that is not parallel to the floor or various other options.

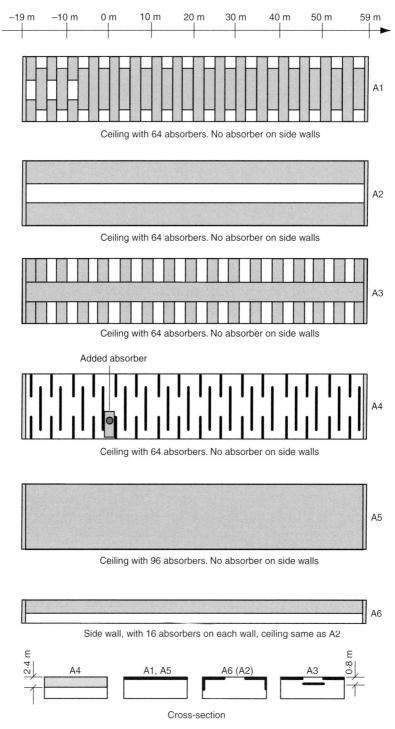

Figure 5.13. *Experimental arrangements of the porous absorbers*

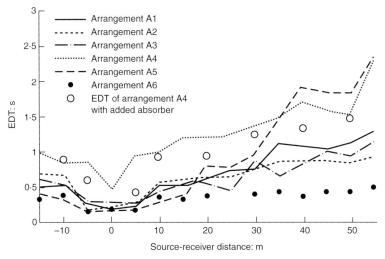

Figure 5.14. EDT along line N with arrangements A1, A2, A3, A4, A5 and A6. Average of 500 Hz and 1000 Hz. Arrangements given in Figure 5.13

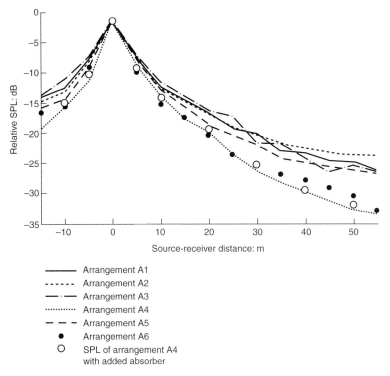

Figure 5.15. Sound attenuation along line N with arrangements A1, A2, A3, A4, A5 and A6. Average of 500 Hz and 1000 Hz. Arrangements given in Figure 5.13

Secondly, arrangements A5 and A6, i.e. the arrangements with 96 absorbers, are compared. The main difference between these two arrangements, as shown in Figure 5.13, is that in arrangement A5 the absorbers are all on the ceiling, whereas in arrangement A6 the absorbers are positioned on both the ceiling and the side walls. From Figures 5.14 and 5.15 it is clear that, when compared with arrangement A5, in arrangement A6 the EDT is much shorter and the sound attenuation along the length is considerably greater. Consequently, the STI from multiple loudspeakers resulting from arrangement A6 is 0·78, which is much higher than that of arrangement A5, i.e. 0·60. This result suggests that the speech intelligibility could be improved by arranging absorbers evenly in cross-section, especially by avoiding multiple reflections between two parallel hard walls. In principle, this corresponds with the theoretical analysis in Section 3.1.

Thirdly, the effectiveness with 64 absorbers is compared to the effectiveness with 96 absorbers. For this comparison, three typical arrangements, namely A1, A5 and A6, are considered. In comparison with arrangement A1, with arrangement A6, where there are 32 additional absorbers on side walls, the EDT is lower and the sound attenuation along the length is greater. Consequently, the STI from multiple loudspeakers is 0·15 higher. Conversely, with arrangement A5, in which the 32 additional absorbers are on the ceiling, the EDT beyond 20 m is considerably longer and the sound attenuation along the length is only slightly greater than that with arrangement A1. As a result, the STI from multiple loudspeakers resulting from arrangement A5 is 0·03 lower than that of arrangement A1. In other words, with more absorbers, the STI from multiple loudspeakers is not necessarily higher.

In conclusion, the above results indicate that the strategic arrangements of absorbers are very effective for improving speech intelligibility. With acoustically smooth boundaries, absorbers are generally more effective when they are evenly distributed in cross-section. With a strategically located absorber arrangement, the STI from multiple loudspeakers could be higher with fewer absorbers.

5.1.6. Membrane absorbers

The model membrane absorbers were arranged from 0 m to 48 m on the ceiling, as illustrated in Figure 5.16. For comparison, measurements were also carried out by arranging the model porous absorbers over the same area.

Figures 5.17 and 5.18 compare the EDT and SPL along line N in three cases, namely with membrane absorbers, with porous absorbers and without absorbers (except for the end walls). It can be seen that with membrane absorbers the EDT is significantly lower and the sound attenuation along the length is systematically greater than that without absorbers. Correspondingly, with membrane absorbers the STI from multiple loudspeakers becomes 0·43

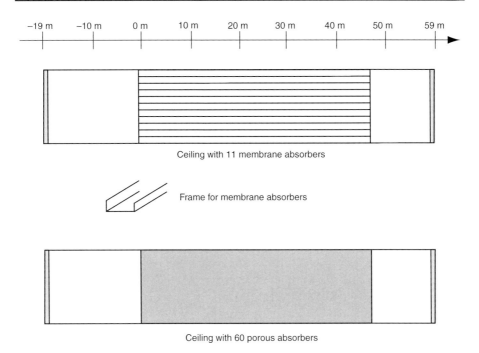

Figure 5.16. Experimental arrangements of the membrane absorber

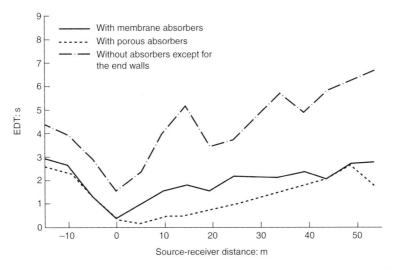

Figure 5.17. Effectiveness of the membrane absorber for the EDT along line N. Average of 500 Hz and 1000 Hz. Arrangements given in Figure 5.16

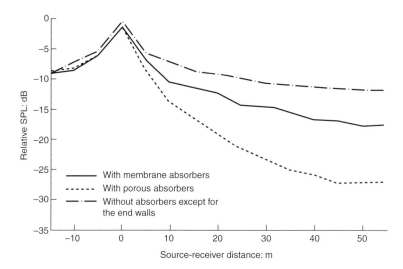

Figure 5.18. Effectiveness of the membrane absorber for the sound attenuation along line N. Average of 500 Hz and 1000 Hz. Arrangements given in Figure 5.16

from 0·20. This result demonstrates that resonant absorbers can be effective for improving speech intelligibility in long enclosures, although the angle of incidence is relatively large.

In comparison with the porous absorber, the membrane absorber is less effective for both the EDT and SPL. Correspondingly, with the porous absorber the STI from multiple loudspeakers is also higher, i.e. 0·59. It appears that these differences are caused mainly by the difference in absorption coefficient between the two absorbers, as shown in Figure 5.2.

To investigate the effect of reflection characteristics of the boundaries, it is useful to compare the EDT and SPL on the left side and right side of the source in the three cases above. The left and right side, as illustrated in Figure 5.1, are from −15 m to 0 m and from 0 m to 50 m, respectively. From Figure 5.18 it can be seen that despite different boundary conditions on the right side, namely with membrane absorbers, with porous absorbers and without absorbers, the SPLs are nearly the same on the left side. This suggests that the reflections on the right side have no significant effect on the SPL on the left. In other words, the boundary reflection appears to be mainly geometrical. By contrast, from Figure 5.17 it can be seen that the EDTs on the left side are noticeably different in the three cases, although the differences are much less than those on the right side. This means that the boundaries are not completely geometrically reflective. Overall, it seems that although geometrical reflection is the main characteristic of the boundaries, diffuse reflection should also be considered. A similar sound field in an actual underground station has been discussed by Barnett [5.15].

5.1.7. Obstructions

The effect of obstructions was investigated with two typical arrangements, as illustrated in Figure 5.19. These arrangements were B1, 16 barriers along the length, and B2, two blocks constructed using 16 barriers. In both cases, 64 absorbers were arranged on the ceiling with the same position as arrangement A2. Figures 5.20 to 5.23 show the SPL and EDT with arrangements A2, B1 and B2. Since the effect of the obstructions is considerably different along lines N and F, the results on both lines are presented.

With arrangement B1 the barriers have opposite effects on the SPL and EDT along the length. In comparison with arrangement A2, in arrangement B1 the SPL is higher within a certain source-receiver distance and lower beyond. This is likely because the barriers reflect certain energy back. Along line F the effectiveness of the barriers are much more significant than that along line N. This effect, which is somewhat similar to that of diffusers, should be positive for improving speech intelligibility. This positive effect, however, is prevented by the negative effect of the barriers on the EDT. From Figures 5.22 and 5.23 it can be seen that in arrangement B1 the EDT is significantly longer on both lines N and F. This is likely to be due to more later reflections caused by the

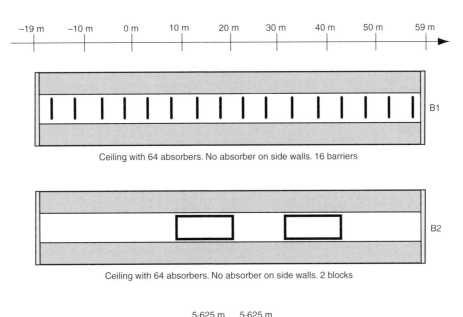

Figure 5.19. Experimental arrangements of the obstructions

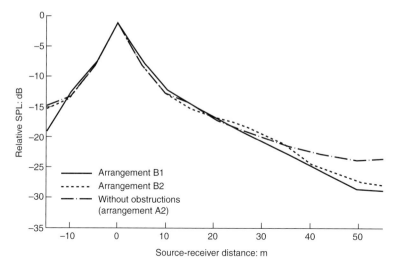

Figure 5.20. Effectiveness of the obstructions for the sound attenuation along line N. Average of 500 Hz and 1000 Hz. Arrangements given in Figure 5.19

barriers. Overall, with arrangement B1 the STI from multiple loudspeakers is 0·55, which is 0·09 lower than that resulting from arrangement A2.

The effect of obstructions is not always negative for the STI. In comparison with arrangement B1, the EDT in arrangement B2 is much shorter. Along line N, the EDT of arrangement B2 is even shorter than that of arrangement A2. A possible reason for this is that the blocks reduce the size of cross-section

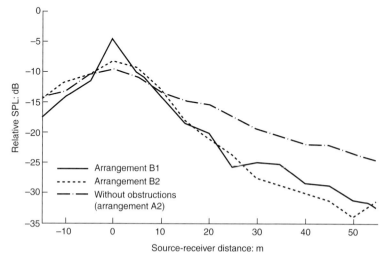

Figure 5.21. Effectiveness of the obstructions for the sound attenuation along line F. Average of 500 Hz and 1000 Hz. Arrangements given in Figure 5.19

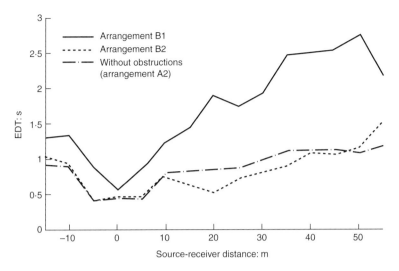

Figure 5.22. Effectiveness of the obstructions for the EDT along line N. Average of 500 Hz and 1000 Hz. Arrangements given in Figure 5.19

and, as a result, later reflections are relatively less. For SPL, the variation with arrangement B2 is similar to that with arrangement B1. Overall, the STI from multiple loudspeakers resulting from arrangement B2 is 0·67, which is higher than that of arrangement A2. It is noted that because the boundary conditions are not the same along the length, the calculation is only an approximate estimation.

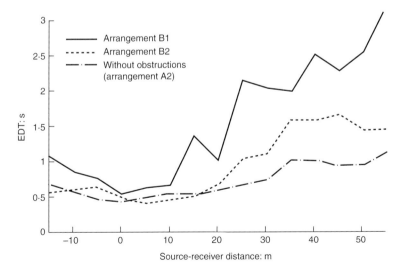

Figure 5.23. Effectiveness of the obstructions for the EDT along line F. Average of 500 Hz and 1000 Hz. Arrangements given in Figure 5.19

In summary, the above examples demonstrate that strategic designs of obstructions can increase the speech intelligibility from multiple loudspeakers.

5.1.8. Further calculations

From the above results it is clear that the STI from multiple loudspeakers is significantly affected by the EDT and the sound attenuation of a single loudspeaker. To investigate this effect in detail, theoretical calculations are made using MUL. The calculations are based on the measurement data of a typical case, arrangement A3. For the sake of convenience, only the loudspeakers on line N are considered (see Figure 5.3). In this case the STI from multiple loudspeakers resulting from arrangement A3 is 0·69.

Firstly, the effect of the EDT from the nearest loudspeaker on the STI from multiple loudspeakers is considered. Generally speaking, the increase of the EDT from the nearest loudspeaker is negative for improving the STI from multiple loudspeakers. For example, with arrangement A3, if the EDT at 0 m is tripled, the STI from multiple loudspeakers becomes 0·63. However, for a shorter EDT of the nearest loudspeaker the early energy is also relatively less and, thus, the EDT from multiple loudspeakers could be longer and the STI from multiple loudspeakers could be lower. In the above case, for example, if the EDT from the nearest loudspeaker is only 1·5 times higher than that of the arrangement A3, the STI from multiple loudspeakers is 0·70, which is even slightly higher than that resulting from arrangement A3.

Secondly, the effect of the EDT from farther loudspeakers on the STI from multiple loudspeakers is considered. In comparison with the nearest loudspeaker, the effect from farther loudspeakers on the STI is relatively less. However, in certain cases this effect could be significant. For example, by tripling the EDT from the loudspeaker at 25 m, the STI from multiple loudspeakers reduces to 0·65. Of course, the effectiveness of a farther loudspeaker is correlated with the SPL attenuation along the length.

Thirdly, the effect of the SPL attenuation along the length on the STI from multiple loudspeakers is considered. Similar to an echo, the high SPL from a farther loudspeaker could seriously decrease the STI from multiple loudspeakers. For instance, with arrangement A3, if the SPL from the loudspeaker at 50 m is 10 dB higher, the STI from multiple loudspeakers becomes 0·61. Using MUL it can be demonstrated that the effect of a farther loudspeaker is unnoticeable when the SPL from this loudspeaker is 20–25 dB lower than that of the nearest loudspeaker.

5.2. A circular sectional long enclosure

In the above Section the effectiveness of a series of architectural acoustic treatments is studied in a rectangular long enclosure with a cross-section of

15 m by 4.8 m. The results, however, are not necessarily applicable when the cross-sectional shape is not rectangular, and the cross-sectional size is relatively small. In this Section, therefore, a series of measurements in a 1:16 scale model of a circular cross-sectional long enclosure is analysed [5.16–5.19].

5.2.1.   The model and modelling technique
5.2.1.1.   Scale model
The 1:16 scale model was a plastic pipe with a length of 8000 mm (128 m full scale) and a diameter of 405 mm (6·48 m full scale). This was a simulation of London St John's Wood underground station, as mentioned in Sections 2.2.1.5, 2.5 and 3.4. The plastic pipe selected for the model (twin wall un-perforated) had external ribs making it suitably rigid and it had a smooth internal bore. The other elements, such as platform, track bed, etc., were made of timber that was varnished with three coats of gloss varnish to give an absorption coefficient of around 0·05.

The model was successfully calibrated against full-scale measurements [5.2,5.3]. For the SPL attenuation along the length, the overall agreement between full scale and model tests was good with all the octave results. Results in the 125 Hz and 250 Hz octave bands fell within 5 dB accuracy except at one point. In the 500 Hz octave band, accuracy was generally within 4 dB of full-scale values. The 1000 Hz octave band was within 3 dB and the 2000 Hz was almost as good. For reverberation, all the model RTs were within 10–15% accuracy with a few exceptions that were marginally outside. For the STI, comparisons of measured values in the model and at full scale showed that full band values were within an accuracy of ±0·05. The repeatability of results in the model, which underwent dismantling and reassembling several times between measurements, was considered to be good.

5.2.1.2.   Model absorbers, diffusers and reflectors
The architectural acoustic treatments tested were porous absorbers, ribbed diffusers, Schroeder diffusers, reflectors and combinations of these treatments.

Two model absorbers, 5 mm thick (8 cm full scale) felt and 10 mm thick (16 cm full scale) plastic foam, were tested. The latter was also used in the rectangular model (see Section 5.1.1). The absorption coefficients of the two absorbers, which were measured in a 1:16 model reverberation room, are shown in Figure 5.24.

The ribbed diffuser described in Section 5.1.1 was also tested in the circular model. In addition, a Schroeder diffuser, also called pseudo-stochastic diffuser or profiled diffuser, which reflects energy equally in all directions, was simulated [5.20–5.22]. The diffusers are periodic surface structures with rigid construction. The elements of the structure are wells of the same width separated by thin fins. Within one period, the depths of the elements vary

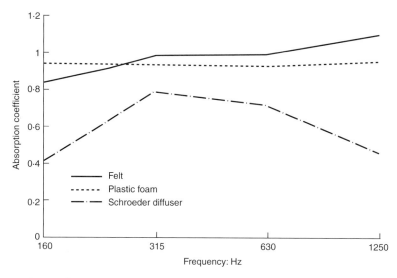

Figure 5.24. Absorption coefficient of the porous absorbers (felt and plastic foam) and the Schroeder diffuser measured in a model reverberation room

according to a pseudo-stochastic sequence. The design frequency of the model Schroeder diffuser was chosen as 10 000 Hz (625 Hz full scale). The maximum frequency and the corresponding well widths were 34 000 Hz (2000 Hz full scale) and 5 mm (8 cm full scale), respectively. Two periods of the pseudo-random sequences with 11 wells per period were chosen, since in this case the diffuser was relatively thin and easy to make. The sequence and the corresponding well depths were (0, 1, 4, 9, 5, 3, 3, 5, 9, 4, 1) and (0, 1·55, 6·18, 19, 7·73, 4·64, 4·64, 7·73, 19, 6·18, 1·55) (mm, model size). The size and thickness of each model Schroeder diffuser were 140 mm by 140 mm (2·24 m by 2·24 m full scale) and 20 mm (32 cm full scale), respectively.

The model Schroeder diffusers were made with thin timber blocks and steel separators. The separators were made as thin as possible (about 1 mm) to avoid the effect of separators themselves. In order to avoid extra absorption, the timber blocks were well varnished and the small gaps between separators and timber blocks were carefully filled with plasticine. However, the absorption of the model Schroeder diffusers was still significant, as shown in Figure 5.24. This might be caused mainly by the well resonance [5.23–5.25].

Since the model Schroeder diffusers are relatively thick and the reduction of cross-section area may affect the results, the effectiveness of the Schroeder diffusers for the SPL variation along the length is measured by the difference in SPL between Schroeder diffusers and plastic-laminated MDF boards with the same size and position.

The model reflectors were made of well-varnished plywood with a thickness of 5 mm (8 cm full scale).

5.2.1.3.   Measurement arrangement

The measurement arrangements are illustrated in Figures 5.25 to 5.27. Measurements were carried out in two categories:

(a)   Architectural acoustic treatments in a section between two adjacent loud-speakers, where the effect of treatments was investigated in detail. For the sake of convenience, it was assumed that the loudspeaker spacing was 6 m.

(b)   Treatments along the length, where the overall effect of multiple sections was investigated.

The measured indices were SPL, EDT and RT30 in octaves. The SPL was measured using a sound level meter, and the EDT and RT30 were measured using a tape recorder with a 10 to 1 speed change facility. To compensate for the excessive air absorption at high frequencies, measurements were carried out using oxygen-free nitrogen in the model at a concentration of 97–99% [5.26, 5.27].

Two sound sources were used. One was a tweeter (Foster Type E120T06) on a side wall, which simulated a PA loudspeaker [5.2]. The other was a sound source (Brüel & Kjær Type HP1001) at the entrance of the tunnel, which simulated the train noise from the tunnel. The receiver was an electret microphone (Knowles Type CA1759) positioned at a height of 94 mm (1·5 m full scale), which corresponded with that of the rectangular model. The distance between receivers and the platform edge was 125 mm (2 m full scale).

5.2.1.4.   A simple method for evaluating results

As analysed previously, the speech intelligibility from multiple loudspeakers can be improved by increasing the SPL from the nearest loudspeaker and, thus, increasing the S/N ratio, and by decreasing the SPL from farther loudspeakers and, thus, decreasing the reverberation from multiple loudspeakers. To evaluate the effectiveness of an architectural acoustic treatment for increasing the SPL from the nearest loudspeaker and decreasing the SPL from a farther loudspeaker in a simple manner, a method is developed as follows.

Assume that between two loudspeakers there are $N_R + 1$ receivers (N_R is an even number) with the same distance between them, as illustrated in Figure 5.28, then the level decrease (LD), level increase (LI) and combined level change (CLC) can be defined as

$$\text{LD} = \frac{2}{N_R} \sum_{i=N_R/2+1}^{N_R} SPL_i \text{ without treatment} - SPL_i \text{ with treatment (dB)}$$

(5.2)

$$\text{LI} = \frac{2}{N_R} \sum_{i=0}^{N_R/2-1} SPL_i \text{ with treatment} - SPL_i \text{ without treatment (dB)}$$

(5.3)

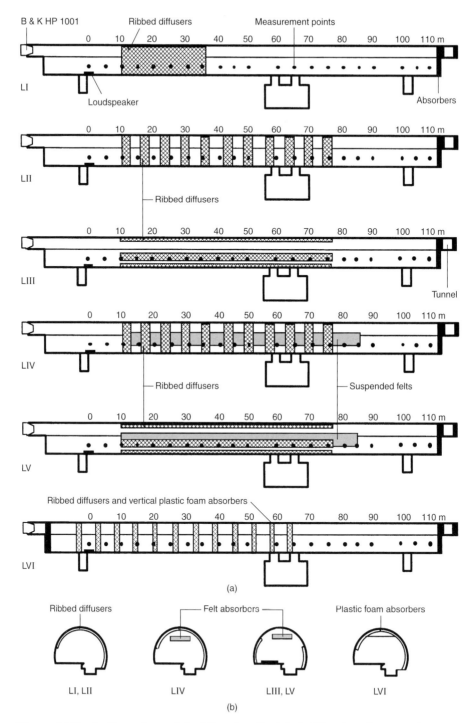

Figure 5.25. *1:16 scale model of London St John's Wood underground station and the measurement arrangements along the model: (a) plan; (b) cross-section*

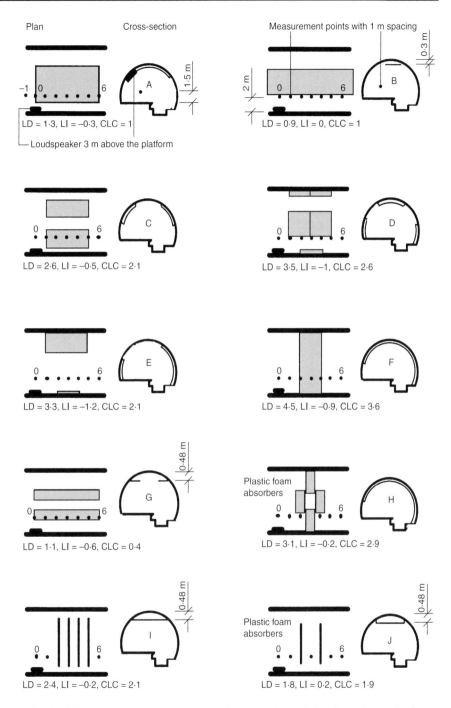

Figure 5.26. Measurement arrangements in a section with felt and plastic foam absorbers (plan and cross-section). The LD, LI and CLC are defined in equations (5.2) to (5.4)

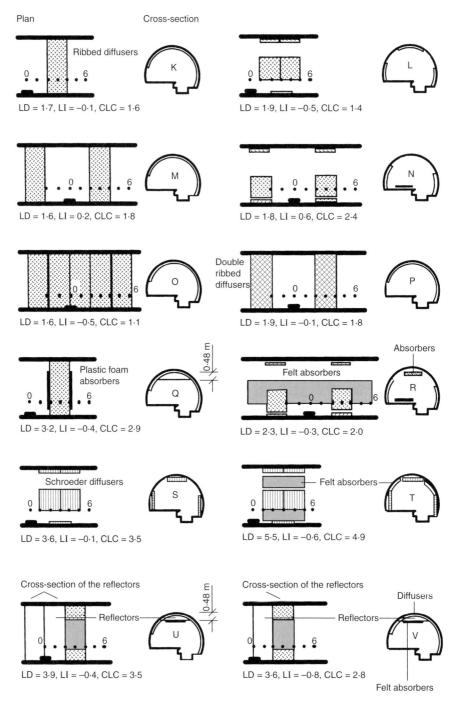

Figure 5.27. Measurement arrangements in a section with absorbers, diffusers and reflectors (plan and cross-section). The LD, LI and CLC are defined in equations (5.2) to (5.4)

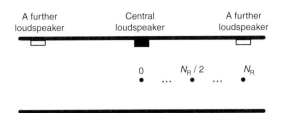

Figure 5.28. Measurement method of the CLC in a long enclosure — plan or length-wise section

$$CLC = LD + LI \quad (dB) \qquad\qquad (5.4)$$

In equations (5.2) and (5.3) the SPL measurements should be carried out using a single loudspeaker, namely the central loudspeaker in Figure 5.28. The LD evaluates the decrease of the SPL from a farther loudspeaker (not necessarily the adjacent one); the LI evaluates the increase of the SPL from the nearest loudspeaker; and the CLC is a combination of the LD and LI. The LD, LI, and CLC can refer to one frequency or the average values of relevant frequencies.

To demonstrate the effect of the CLC on the STI, a simple calculation using MUL (see Section 2.6.4) is carried out under the following conditions (see Figure 5.28):

(*a*)   the aim point is $i = 0$;
(*b*)   two farther loudspeakers with the same characteristics as the central loud-speaker are considered;
(*c*)   at receiver 0, the SPL of the central loudspeaker is 5 dB higher than that of receiver N_R;
(*d*)   the EDT of the central loudspeaker is 0·5 s at receiver 0 and 1·5 s at receiver N_R;
(*e*)   the S/N ratio is 30 dB; and
(*f*)   for a given CLC there are equal SPL decreases at receivers $(N_R/2) + 1$ to N_R and equal increases at receivers 0 to $(N_R/2) - 1$.

The calculation results are shown in Figure 5.29. It can be seen that the effect of the CLC on the STI is significant.

The CLC between two adjacent loudspeakers has the greatest importance for the STI. For a given loudspeaker spacing, this CLC is more sensitive to a strategic architectural acoustic treatment on the boundaries between the two loudspeakers when the cross-section is relatively small. For this reason, the CLC is used for this circular sectional model to compare various architectural acoustic treatments.

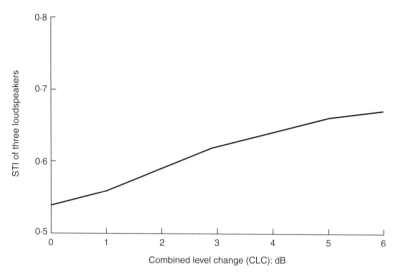

*Figure 5.29. Effect of the CLC on the STI from three loudspeakers — calcula-
tion arrangement given in Figure 5.28*

5.2.2. Treatments in one section

The following analyses focus on the CLC between two adjacent loudspeakers,
the reverberation time of a single loudspeaker and the STI from multiple loudspeakers. The CLCs with various treatments in one section are shown in
Figures 5.26 and 5.27. In a similar manner to the CLC, the extra SPL attenuation along the length caused by a treatment is also used to compare various
treatments. The CLC and extra attenuation below are the arithmetical
means of 500 Hz, 1000 Hz and 2000 Hz, except where indicated.

5.2.2.1. Absorbers

To investigate the effect of the distribution of absorption on the CLC, a series
of measurements was carried out with a given amount of absorption but different absorber distributions, as illustrated in Figure 5.26. The results with the felt
absorbers (arrangements A to F) are shown in Figure 5.30. It can be seen that
the CLC is the lowest (about 1 dB) when the absorbers are along the ceiling and
the highest (about 4 dB) when the absorbers are between 2 m and 4 m. Similar
phenomena can also be seen in the measurements using the plastic foam
absorbers. In Figure 5.31, for example, it is shown that in comparison with
the absorbers along the ceiling (arrangement G), when the absorbers are
between 2 m and 4 m (arrangement H), the extra attenuation is significantly
higher. The main reason for this is that with various absorber arrangements
the chances of absorbing the reflecting energy are different.

Correspondingly, the distribution of absorption also affects reverberation.
Figure 5.32 shows the RT30 and EDT at 630 Hz corresponding to Figure 5.31.

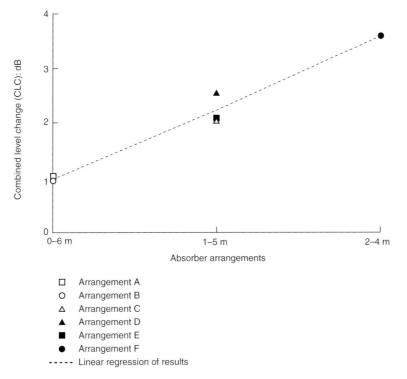

Figure 5.30. CLCs with a given amount of absorption but different absorber distributions. Arrangements given in Figure 5.26

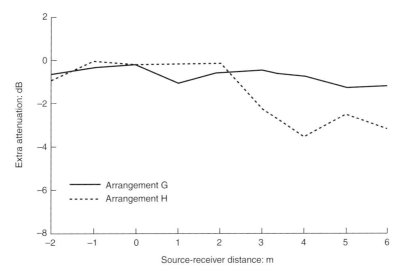

Figure 5.31. Extra attenuation with a given amount of absorption but different absorber distributions. Average of 500 Hz, 1000 Hz and 2000 Hz. Arrangements given in Figure 5.26

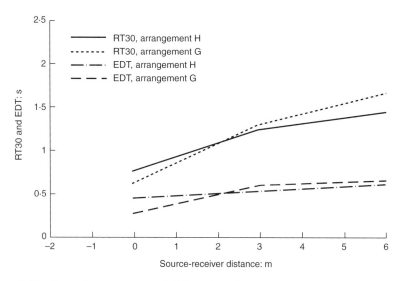

Figure 5.32. Reverberation time (630 Hz) with a given amount of absorption but different absorber distributions. Arrangements given in Figure 5.26

In comparison with the absorbers along the ceiling, when the absorbers are between 2 m and 4 m, the RT30 and EDT are longer at 0 m and shorter beyond 3 m. A possible reason, which is similar to that for the CLC, is that the absorbers between 2 m and 4 m are more effective for absorbing reflections beyond 3 m. Since the EDT beyond 3 m is longer than that at 0 m, the decrease of the EDT beyond 3 m could result in a more significant benefit for the STI and the STI distribution along the length could be more consistent.

By considering both the CLC and reverberation time, when the absorbers are between 2 m and 4 m the average STI from multiple loudspeakers should be higher than that of absorbers along the ceiling.

The effect of the amount of absorption was also investigated. By comparing arrangements I and J (see Figure 5.26) it is interesting to note that with many fewer absorbers the CLC could be nearly the same. This strategic arrangement, however, appears to be less effective for reverberation. In Figure 5.33 it is shown that with arrangement I the RT30 and EDT are systematically shorter than those with arrangement J. A possible reason for this phenomenon is that the CLC can be changed by varying early reflections but the reverberation depends mainly on multiple reflections.

With a given amount and distribution of absorption, the absorption coefficient also affects the CLC. For example, with arrangement D the CLCs of the plastic foam absorber and the felt absorber are 2·2 dB and 2·6 dB, respectively. This difference is likely to be caused by the difference in absorption coefficient, as shown in Figure 5.24.

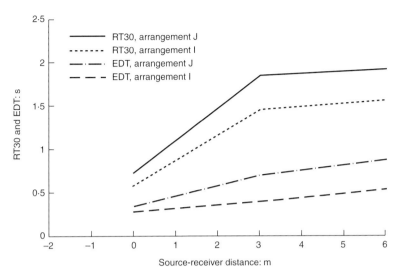

Figure 5.33. Reverberation time (630 Hz) with different amount of absorption. Arrangements given in Figure 5.26

5.2.2.2. Ribbed diffusers

The extra attenuation caused by the ribbed diffusers with arrangements K and L (see Figure 5.27), where there is a single group of diffusers, is shown in Figure 5.34. It can be seen that with diffusers the SPL is slightly higher between −2 m and 1 m and significantly lower beyond 2 m. This corresponds with the

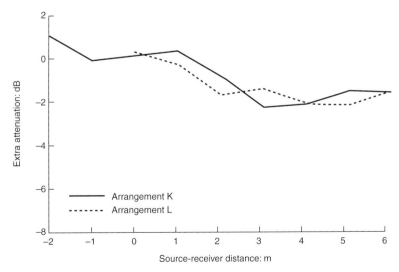

Figure 5.34. Extra attenuation by the ribbed diffusers with arrangement K and arrangement L. Average of 500 Hz, 1000 Hz and 2000 Hz. Arrangements given in Figure 5.27

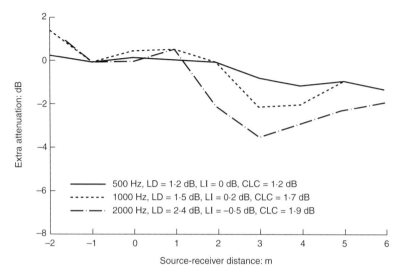

Figure 5.35. Extra attenuation by the ribbed diffusers with arrangement K. Arrangement given in Figure 5.27

theoretical analysis in Section 3.1 and the measurement results in Section 5.1.4. As expected, from 500 Hz to 2000 Hz, the CLC increases with the increase in frequency. The extra attenuation and CLC at the three frequencies with arrangement K, for example, is shown in Figure 5.35.

As with arrangements K and L, Figure 5.36 shows the extra attenuation with arrangements M and N, in which there is a group of diffusers on both sides of the source. In comparison with the results in Figure 5.34, the SPL is systematically higher between −1 m and 3 m and nearly the same beyond 4 m. This indicates the effective area of the diffusers on the left side (see Figure 5.27).

The extra attenuation with arrangement O, in which there are five groups of diffusers between −4 m and 6 m, is also illustrated in Figure 5.36. In comparison with arrangements M and N, with arrangement O the SPL between the source and 2 m is systematically lower and not higher beyond 3 m. This means that with more diffusers the total energy is lower, which is theoretically reasonable (see Section 2.1.1). Correspondingly, with arrangement O the CLC is lower due to the low LI, as shown in Figure 5.27. From the viewpoint of speech intelligibility of multiple loudspeaker PA systems, arrangements M or N are slightly better than arrangement O, since their SPL between 1 m and −1 m is higher. From the viewpoint of noise reduction, arrangement O is better.

In Figure 5.36 it can be seen that with arrangements M and N, the SPLs at 2 m and 3 m are rather different. This means that the SPL distribution could be adjusted, more or less, by the arrangement of diffusers.

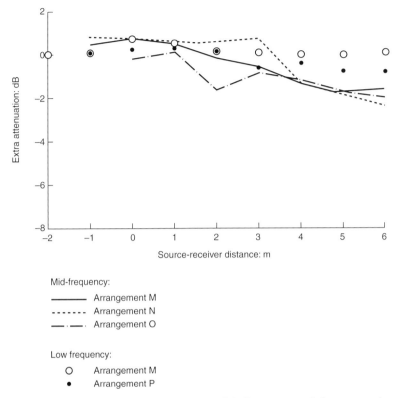

Figure 5.36. Extra attenuation by the ribbed diffusers at mid-frequency (average of 500 Hz, 1000 Hz and 2000 Hz) and at low frequency (125 Hz). Arrangements given in Figure 5.27

As expected, the effectiveness of the ribbed diffusers is unnoticeable at 125 Hz (see Section 5.1.1). The extra attenuation at this frequency with arrangement M, for example, is shown in Figure 5.36. This effectiveness, however, could be improved by increasing the diffuser thickness. Figure 5.36 also shows the extra attenuation at 125 Hz with arrangement P, in which the diffuser position is the same as that with arrangement M but the diffuser thickness is doubled. It can be seen that with the increase of thickness, the extra attenuation at this frequency increases correspondingly.

It is interesting to note that the ribbed diffusers are also effective for decreasing reverberation in this circular sectional model. Figure 5.37, in correspondence with Figure 5.35, shows a comparison of RT30 and EDT at 630 Hz with and without ribbed diffusers. It can be seen that the RT30 and EDT are decreased systematically by the diffusers. The average decreases of the RT30 and EDT at the seven receivers are 33% and 17%, respectively. A possible reason for this decrease is that with these diffusers the decay

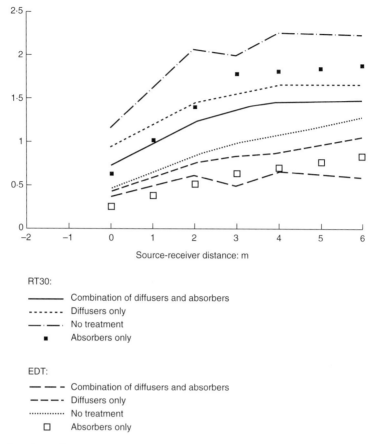

Figure 5.37. Reverberation time (630 Hz) with the ribbed diffusers and absorbers in arrangement Q. Arrangement given in Figure 5.27

curves are more linear. This can be demonstrated by the fact that with diffusers the RT30/EDT ratio is systematically lower, although still greater than 1. Figure 5.38 shows the RT30/EDT ratio in correspondence with Figure 5.37.

5.2.2.3. Combination of ribbed diffusers and absorbers

The extra attenuation caused by the ribbed diffusers with arrangements Q and R (see Figure 5.27), in which there are suspended plastic foam absorbers, is shown in Figure 5.39. It can be seen that the diffusers are still effective under highly absorbent conditions.

The extra attenuation caused by the combination of ribbed diffusers and absorbers in arrangement Q is shown in Figure 5.40. In comparison with absorbers (arrangement J) only or ribbed diffusers only (arrangement K), the CLC of the combination is greater than either and less than their sum.

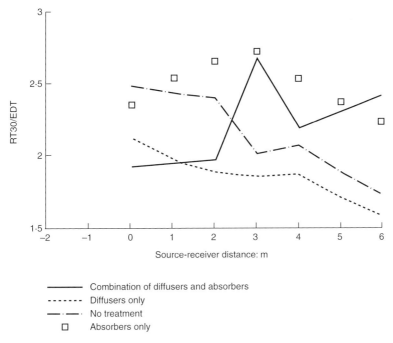

Figure 5.38. RT30/EDT ratio (630 Hz) with the ribbed diffusers and absorbers in arrangement Q. Arrangement given in Figure 5.27

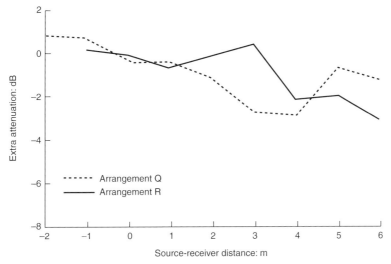

Figure 5.39. Extra attenuation caused by the ribbed diffusers under highly absorbent conditions. Average of 500 Hz, 1000 Hz and 2000 Hz. Arrangements given in Figure 5.27

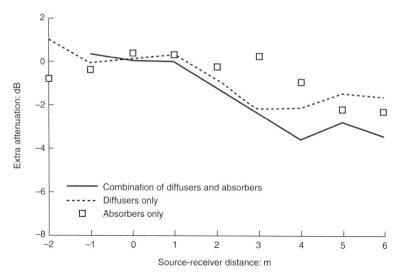

Figure 5.40. Extra attenuation by the ribbed diffusers and absorbers with arrangement Q. Average of 500 Hz, 1000 Hz and 2000 Hz. Arrangements given in Figure 5.27

The RT30 and EDT of the combination are shown in Figure 5.37. In comparison with ribbed diffusers only, the RT30 and EDT of the combination are systematically shorter. In comparison with absorbers only, the RT30 and EDT of the combination are longer between 0 m and 2 m and shorter beyond. Possibly, this is because the ribbed diffusers between 2 m and 4 m increase the reflection energy between 0 m and 2 m and decrease the reflection energy beyond. The RT30/EDT ratio of the combination (as well as absorbers only) is generally greater than that of ribbed diffusers only, as shown in Figure 5.38.

By considering both the CLC and reverberation time, the STI of the combination is better than either ribbed diffusers or absorbers.

5.2.2.4. Schroeder diffusers

The extra attenuation caused by a group of Schroeder diffusers with arrangement S (see Figure 5.27) is shown in Figure 5.41. To reduce the effect of cross-sectional change, these diffusers were arranged between 1 m and 5 m. As expected, the extra attenuation at 500 Hz is lower than that at 1000 Hz and 2000 Hz, since the design frequency of the Schroeder diffusers was above 625 Hz. In comparison with the felt absorbers with the same arrangement (arrangement D), as shown in Figure 5.42, the CLC of the Schroeder diffusers is 1 dB higher, which means that diffusion still plays an important role for the CLC since the absorption coefficient of the Schroeder diffusers is much lower than that of the felt absorbers (see Figure 5.24). In comparison with the ribbed

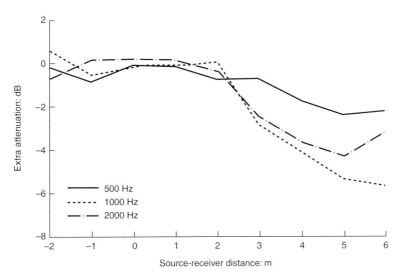

Figure 5.41. Extra attenuation by the Schroeder diffusers with arrangement S. Arrangement given in Figure 5.27

diffusers with the same arrangement (arrangement L), the CLC of the Schroeder diffusers is 2·2 dB higher. This is likely due to the optimal diffusing characteristic and extra absorption of the Schroeder diffuser. Similar to the ribbed diffusers, the Schroeder diffusers are also very effective for decreasing reverberation, as shown in Figure 5.43.

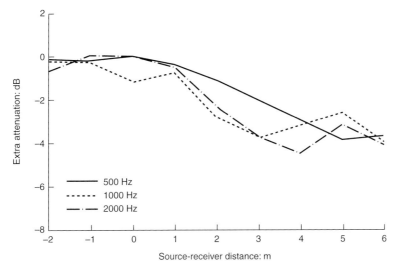

Figure 5.42. Extra attenuation by the felt absorbers with the same arrangement (arrangement D) as the Schroeder diffusers in Figure 5.41. Arrangements given in Figure 5.26

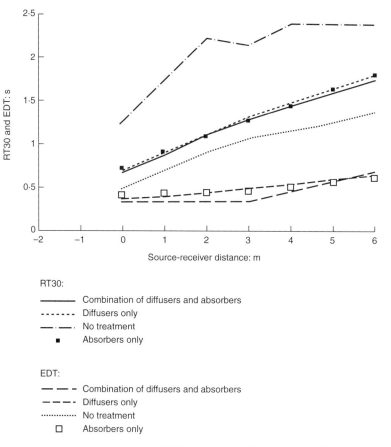

Figure 5.43. Reverberation time (630 Hz) with the Schroeder diffusers and absorbers in arrangement T. Arrangement given in Figure 5.27

5.2.2.5. Combination of Schroeder diffusers and absorbers

The extra attenuation caused by the combination of Schroeder diffusers and felt absorbers in arrangement T is shown in Figure 5.44. In comparison with absorbers only or Schroeder diffusers only, the CLC of the combination is greater than either and slightly lower than their sum. In other words, the Schroeder diffusers are still effective under a highly absorbent condition. This corresponds with the results of ribbed diffusers (see Figure 5.39).

Conversely, the RT30 and EDT of the combination is nearly the same as, or only slightly shorter than, that of absorbers or Schroeder diffusers alone, as shown in Figure 5.43. A possible reason is that the efficiency of extra absorption becomes lower when the absolute reverberation time becomes shorter. By considering both the CLC and reverberation, it appears that for the STI, with the Schroeder diffusers the benefit of extra absorbers is not significant in the above case.

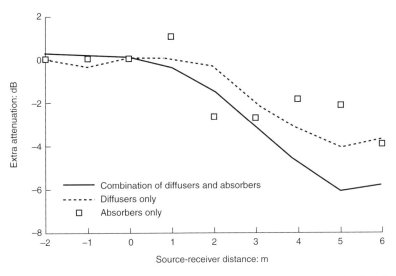

Figure 5.44. Extra attenuation by the Schroeder diffusers and absorbers in arrangement T. Average of 500 Hz, 1000 Hz and 2000 Hz. Arrangement given in Figure 5.27

5.2.2.6. Reflectors

The CLC can also be increased by changing the reflection pattern through sound reflectors. Figure 5.45 shows an example with two reflector arrangements, from −2 m to 2 m (arrangement U) and from 0 m to 2 m (arrangement

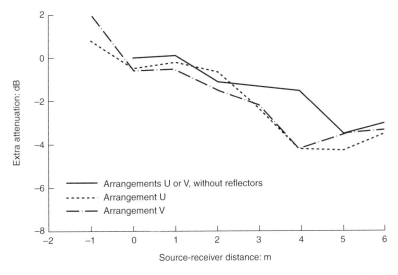

Figure 5.45. Extra attenuation by the ribbed diffusers, absorbers and reflectors. Average of 500 Hz, 1000 Hz and 2000 Hz. Arrangements given in Figure 5.27

V), where there are absorbers and ribbed diffusers between 2 m and 4 m. It can be seen that with the reflectors between −2 m and 2 m, the CLC is about 1·3 dB higher than that without reflectors. When the reflectors are only between 0 m and 2 m, the CLC, especially the LI, is lower than that of reflectors between −2 m and 2 m. Possibly this is because some sound energy is reflected to the area beyond 0 m.

5.2.3.   Treatments along the length
5.2.3.1.   Ribbed diffusers

The extra attenuation caused by the ribbed diffusers with arrangement LI (see Figure 5.25) is shown in Figure 5.46, where the source is the loudspeaker. The data are the arithmetical means of 500 Hz and 1000 Hz. It can be seen that from the source to 15 m the SPL increases slightly, about 1 dB, and at greater distances the extra attenuation is significant. This corresponds with the results in one section. The maximum extra attenuation occurs between 40 m and 60 m, which is the area just after the diffusers. This is because in this area the sound energy is decreased significantly by the diffusers but not increased by the boundaries beyond. The maximum extra attenuation is about 6 dB, which is significant in comparison with the absolute sound attenuation along the length. When there was no diffuser, the attenuation from 0 m to 50 m was 16 dB at 500 Hz.

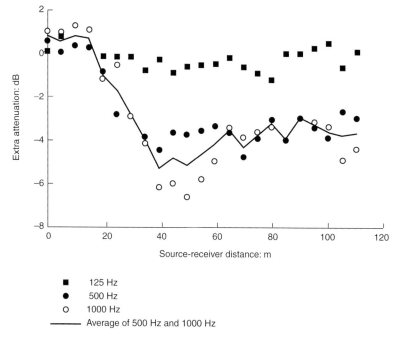

Figure 5.46. Extra attenuation by the ribbed diffusers in arrangement LI with the loudspeaker source. Arrangement given in Figure 5.25

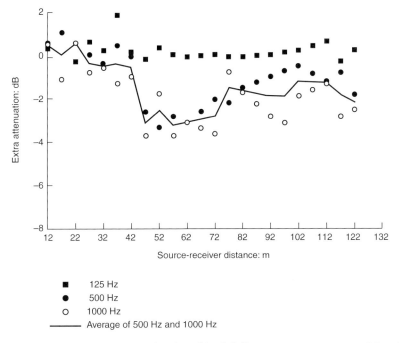

Figure 5.47. *Extra attenuation by the ribbed diffusers in arrangement LI with the train source (B&K HP1001). Arrangement given in Figure 5.25*

Corresponding to Figure 5.46, the extra attenuation with the train source (Brüel & Kjær HP1001) is shown in Figure 5.47. The form of the plot is similar to that for the loudspeaker but the extra attenuation is about 2–3 dB less. A possible reason is that with the train source the direct sound plays a more important role and the effect of the diffusers is thus diminished.

As expected, the effect of the diffusers at 1000 Hz is greater than that at 500 Hz (see Figures 5.46 and 5.47), especially between 40 m and 60 m. The extra attenuation at 125 Hz is close to zero. This is similar to the results in one section.

The extra attenuation with arrangements LII and LIII (see Figure 5.25) is shown in Figure 5.48, where the source is the loudspeaker. In comparison with arrangement LI, the extra attenuation between 40 m and 60 m is about 2 dB less. This is partly because the diffusers before this area (i.e. 10–40 m) are less effective, due to the low diffuser density, and partly because the sound energy in this area is increased by the diffusers beyond (i.e. >60 m).

5.2.3.2. Combination of ribbed diffusers and absorbers

The extra attenuation caused by the ribbed diffusers with arrangements LIV and LV (see Figure 5.25), where the diffuser distributions are the same as arrangements LII and LIII but there are suspended felt absorbers along the length, is shown in Figure 5.49. It can be seen that, similar to the results in

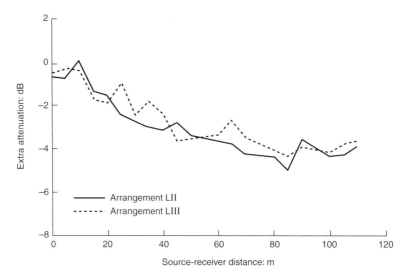

Figure 5.48. Extra attenuation by the ribbed diffusers in arrangements LII and LIII with the loudspeaker source. Average of 500 Hz and 1000 Hz. Arrangements given in Figure 5.25

one section, the diffusers are also effective under highly absorbent conditions. In comparison with Figure 5.48, the extra attenuation in Figure 5.49 is slightly less. This corresponds with the results in Section 5.1.4.

Between arrangements LII and LIII, and between LIV and LV, generally speaking, there is no significant difference in extra attenuation, as shown in

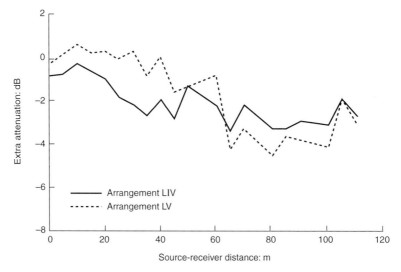

Figure 5.49. Extra attenuation by the ribbed diffusers in arrangements LIV and LV with the loudspeaker source. Average of 500 Hz and 1000 Hz. Arrangements given in Figure 5.25

Figures 5.48 and 5.49. The slight difference between the two arrangements in Figure 5.49 might be caused by the fact that, with arrangement LIV, some diffusers were hindered by the absorbers.

5.2.3.3. STI from multiple loudspeakers

To demonstrate the effectiveness of architectural acoustic treatments on the speech intelligibility of multiple loudspeaker PA systems and to investigate the overall effect of multiple sections, a test was carried out with a treatment along about 60 m. The treatment, as illustrated in Figure 5.25 (arrangement LVI), is a combination of ribbed diffusers and plastic foam absorbers. This corresponds with arrangement Q in one section (see Figure 5.27). Figures 5.50 and 5.51 show the measured EDT and SPL along the length at 630 Hz, both with and without the treatment. It can be seen that the treatment has a significant effect on both the EDT and SPL.

The STI from multiple loudspeakers is calculated using MUL. In the calculation the loudspeaker spacing is 6 m and the S/N ratio is 30 dB(A). The results show that, if there are 10 loudspeakers on each side of a central loudspeaker, by using the above treatment along the whole station the STI from multiple loudspeakers at a receiver in the same cross-section as the central loudspeaker becomes 0·71 from 0·43. At 3 m from the central loudspeaker, namely, halfway between two loudspeakers, the STI from multiple loudspeakers becomes 0·55 from 0·37. This improvement is significant.

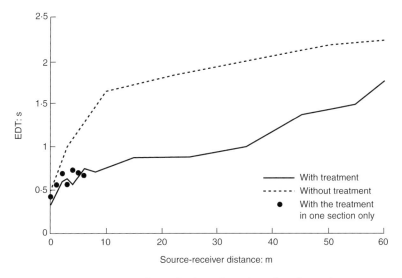

Figure 5.50. EDT (630 Hz) along the length with and without the treatment (see arrangement LVI in Figure 5.25). Also shown is the EDT with the treatment in one section only (see arrangement Q in Figure 5.27)

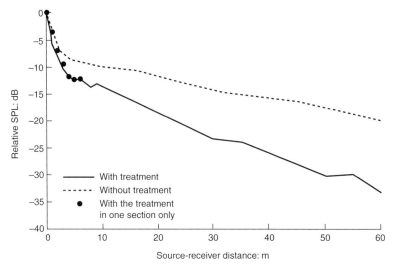

Figure 5.51. SPL (630 Hz) along the length with and without the treatment (see arrangement LVI in Figure 5.25). Also shown is the SPL with the treatment in one section only (see arrangement Q in Figure 5.27)

Figures 5.50 and 5.51 also compare the EDT and SPL of the same treatment in one and multiple sections. The EDT with one section treatment is slightly longer than that of multiple section treatments. This suggests that the treatments in the adjacent sections are helpful for absorbing later reflections. Conversely, the difference in SPL between one section and multiple section treatments is much less. Possibly this is because the SPL depends mainly on early reflections and, thus, the effect of other sections is less important. In general, the results with a treatment in one section correspond to those with the same treatment in multiple sections. In other words, the above analyses and comparisons in one section are also applicable when the treatments are along the length of the station.

5.2.4. Train noise
First, this Section presents a series of scale model measurements for determining the database of a train section source. The database is essential for TNS (see Section 2.5). Particular attention is given to the effect of the whole train on the database when the train section source is at a different position of the train and when the train length varies. Using TNS, this Section then analyses the distribution of train noise in the station and the effectiveness of architectural acoustic treatments on reducing train noise [5.28,5.29].

5.2.4.1. Model train section and model train
The measurement arrangements for train noise are illustrated in Figure 5.52. The boundary conditions were approximately the same along the station. On either end of the model, a short tunnel was simulated.

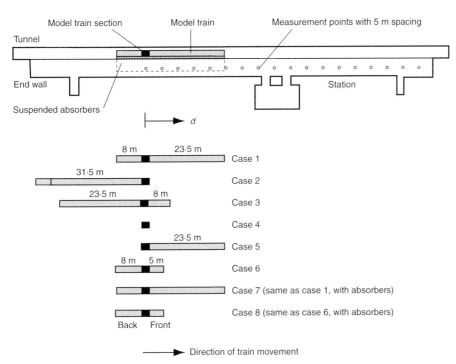

Figure 5.52. Measurement arrangements for train noise — plan view

The model train section and the model train are illustrated in Figure 5.53. The model train section was a plastic pipe with a diameter of 160 mm (2·56 m full scale). The thickness of the pipe was 6 mm. A tweeter (Foster Type E120T06) was put underneath the model section so that the centre of the source was close to the wheel and, thus, simulated the wheel/rail noise. Two 15 mm thick plastic laminated MDF boards with approximately the same shape as the actual train section were positioned on either end of the model train section. The model train was also made of the same pipe as the model train section, with two plastic laminated MDF boards (15 mm thick) of approximately the same shape as the actual train façade on either side. The model train was in a number of sections so that the model train section (i.e. the sound source) can be put in the required position.

5.2.4.2. Effect of train on Sss

As mentioned in Section 2.5.2, *Sss* represents the sound attenuation from a train section to a receiver when they are both in the station. To investigate the variation of *Sss* when a train section is at different positions in a train, measurements were made in the scale model with eight cases, as shown in Figure 5.52.

A comparison of the SPL attenuation at 500 Hz with and without the model train on two sides of the model train section (cases 1, 3 and 4) is shown in

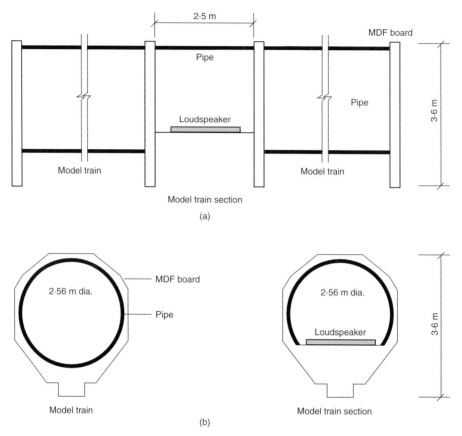

Figure 5.53. Model train section source and model train: (a) length-wise section; (b) cross-section

Figure 5.54. It shows that with the model train the SPL is lower. A possible reason is that the model train prevents the noise radiation of the model train section source from the opposite side of the platform.

Figure 5.55 shows a comparison of the SPL at 500 Hz with and without the model train behind the model train section when there was a model train in front of the model train section (cases 1 and 5). It can be seen that the difference is systematic, about 2–3 dB. This indicates that in this model the radiation of the model train section in the direction of the back cab also has an effect on the direction of the front cab. This effect was less when there was no model train in front of the model train section, which is shown in Figure 5.56 (cases 2 and 4). This is because, in this case, the radiation from the model train section in the front direction is relatively strong.

A comparison between two train lengths (cases 1 and 6) is shown in Figure 5.57. It can be seen that with a longer model train the SPL is systematically

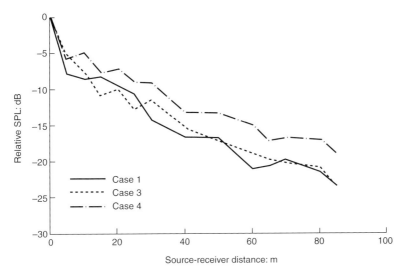

Figure 5.54. Scale model measurement of the train's effect on Sss (500 Hz) with and without the model train on two sides of the model train section

lower, especially beyond the model train in case 6. In principle, this corresponds with the results in Figure 5.54.

To investigate the effect of the train on *Sss* under different absorbent conditions in the station, the comparison in Figure 5.57 was also made with suspended felt absorbers on part of the ceiling (see Figure 5.52). Figure 5.58

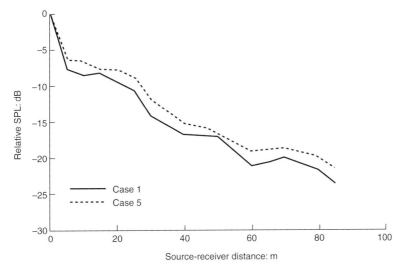

Figure 5.55. Scale model measurement of the train's effect on Sss (500 Hz) with and without the model train behind the model train section when there is a model train in front of the model train section

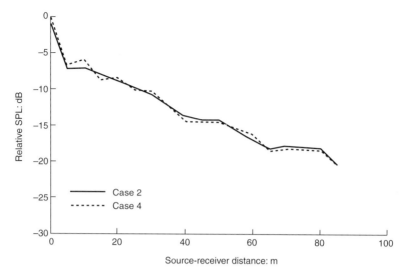

Figure 5.56. Scale model measurement of the train's effect on Sss (500 Hz) with and without the model train behind the model train section when there is no model train in front of the model train section

shows that in this case the SPL is also lower with a longer model train, especially beyond the model train in case 8. This is similar to the results in Figure 5.57.

In summary, the above analyses demonstrate that in this station the model train has a systematic effect on *Sss*. The difference in the SPL at middle

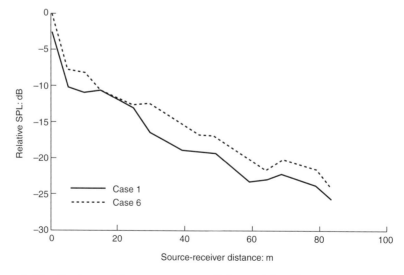

Figure 5.57. Scale model measurement of the train's effect on Sss (500 Hz): a comparison between two train lengths

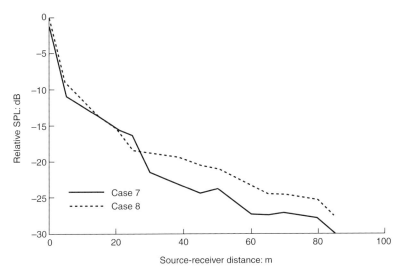

Figure 5.58. Scale model measurement of the train's effect on Sss (500 Hz): a comparison between two train lengths. The configuration corresponds to that in Figure 5.57 but there were suspended absorbers on part of the ceiling

frequencies with and without a train is about 5 dB and its variation over different positions of the train section source in the train is 2–3 dB.

5.2.4.3. Effect of train on Sts

Sts represents the relative SPL attenuation from the tunnel entrance to the station (see Section 2.5.2). The effect of the train on *Sts* was measured using the sound source Brüel & Kjær HP1001, which has a similar size to the tunnel cross-section. The measurement arrangements are illustrated in Figure 5.59. Three cases, namely, with a long model train, with a short model train and without a model train, were considered. The comparison of *Sts* at 500 Hz in these three cases is shown in Figure 5.60. It can be seen that within about 30 m from the tunnel entrance, which is approximately the range of the long model train, the SPL with the long model train is systematically lower than that with the short model train. This corresponds to the results of *Sss*, as shown in Figures 5.57 and 5.58. Beyond this range, conversely, the difference in SPL between the two train lengths is not noticeable. This is different from the results in Figures 5.57 and 5.58, which might be caused by the different source characteristics. As expected, the difference in SPL with and without the model train is systematic, at about 5 dB.

5.2.4.4. Train noise in station

The calculation below is carried out using the following parameters (also see Section 2.5):

(a) $L_T = 75$ m, $N_T = 30$ and $D_s = 128$ m;

(b) three Sss (cases 1, 3 and 6 in Figure 5.52);

(c) three Sts (cases A, B and C in Figure 5.59);

(d) $Stt(d) = Stt(d - 1) - 0.1$ for $d > 60$ m and $Stt(d) = Sss(1.5d)$ otherwise (theoretical estimation);

(e) $S(n, v) = S(1, 60) + 10\log(v/60)$ and $S(n, 0) = S(n, 60) - 13$ (based on equation (2.95) and site measurement);

(f) uniform deceleration and acceleration with $V_d = V_i = 54$ km/h; and

(g) $T_a = 10$ s, $T_d = 20$ s, $T_s = 10$ s, $T_i = 20$ s and $T_e = 10$ s.

Figure 5.61 shows the temporal train noise distribution at 500 Hz at three points with $r = 10$ m, 35 m and 60 m. In comparison with 35 m and 60 m, at 10 m the SPL is higher before the train enters the station and much lower after the train passes this point. In other words, the overall effect of train noise on the area near the end wall where the train enters the station is slightly less than that on the other areas. Calculations at $r = 70$ m, 90 m and 110 m show that this conclusion is also applicable to the other end wall.

Figure 5.62 shows the spatial train noise distribution at 500 Hz at three specified times: 10 s, the train is in the tunnel; 20 s, the train is decreasing in

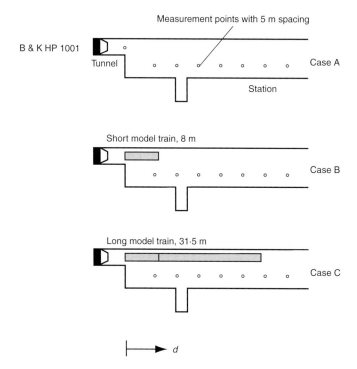

Figure 5.59. Measurement arrangements for investigating the train's effect on Sts in the scale model — plan view

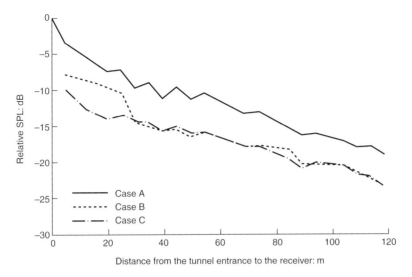

Figure 5.60. Scale model measurement of the train's effect on Sts (500 Hz)

velocity; and 35 s, the train is static. It can be seen that at the three times above the spatial distributions are significantly different.

The effect of train noise on the STI of a multiple loudspeaker PA system is demonstrated in Figure 5.63, where the receiver is at $r = 60$ m. In the calculation it is assumed that the loudspeaker spacing is 5 m and the signal level from the nearest loudspeaker is 80 dB(A). It can be seen that due to the effect of train noise, there are two obvious troughs on the STI curve.

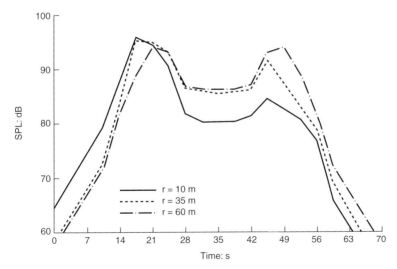

Figure 5.61. Temporal train noise distribution at three receivers (500 Hz), calculated using TNS

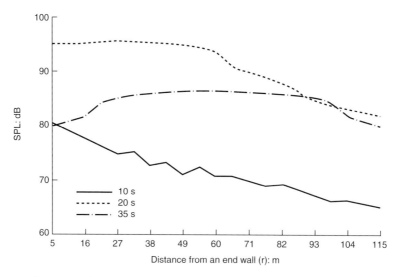

Figure 5.62. Spatial train noise distribution at three times (500 Hz), calculated using TNS

To investigate the effectiveness of some conventional architectural acoustic-treatments for reducing train noise, calculations are also made by considering:

(*a*)   suspended absorbers on the ceiling, where it is assumed that in *Sss* and *Sts* there is a 0·1 dB/m extra attenuation at 500 Hz; and

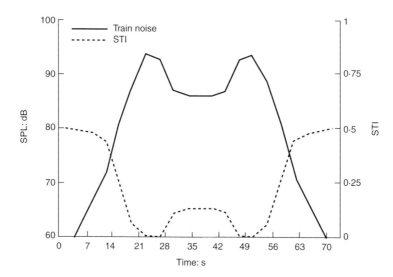

Figure 5.63. Effect of train noise on the STI of a multiple loudspeaker PA system, calculated using TNS — r = 60 m

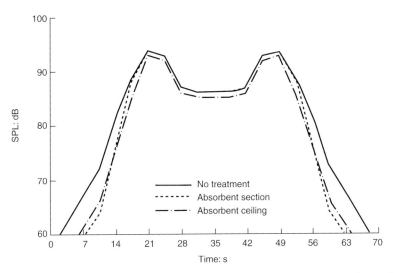

SPL: dB

Time: s

Figure 5.64. Effectiveness of two architectural acoustic treatments for reducing train noise (500 Hz, r = 60 m), calculated using TNS

(*b*)   strong absorbent sections near the two ends, where it is assumed that there is an 8 dB extra attenuation in *Sts* at 500 Hz.

These assumptions are based on the measurements in the scale model.

The effectiveness of the above two treatments at 500 Hz is shown in Figure 5.64, where *r* = 60 m. It is seen that the treatments are effective when the train is in the tunnel but are not as helpful in reducing train noise when the train is in the station. The reason is that at this time, direct sound is more significant than reflections, given the fact that treatments are only effective on the latter.

5.3. Summary

The physical scale modelling, which particularly relates to two typical underground stations, has demonstrated the effectiveness of strategic architectural acoustic treatments in long enclosures, especially for improving the speech intelligibility of multiple loudspeaker PA systems:

(*a*)   The efficiency of absorbers can be affected significantly by their arrangements. With a strategically located absorber arrangement, the intelligibility from multiple loudspeakers could be higher with fewer absorbers.

(*b*)   Some absorber arrangements, which are very effective for increasing the sound attenuation along the length, are not necessarily effective for decreasing reverberation.

(c)   Absorbent treatments on end walls are very useful for improving the speech intelligibility in long enclosures.

(d)   For a relatively small cross-section, when absorbers are located halfway between two loudspeakers, the average STI from multiple loudspeakers is higher than that with absorbers along the ceiling.

(e)   Despite the relatively large angle of sound incidence in long enclosures, resonant absorbers, such as a membrane absorber, can still be effective for increasing the sound attenuation along the length and for decreasing reverberation and, thus, increasing the speech intelligibility of multiple loudspeaker PA systems.

(f)   Ribbed diffusers are effective for increasing the sound attenuation along the length, for both train noise and loudspeaker sources. They are useful under conditions of both high and low absorption but the effectiveness generally becomes less with increasing boundary absorption.

(g)   With a relatively small cross-section, ribbed diffusers can be effective for increasing the SPL from the nearest loudspeaker and for decreasing the SPL from farther loudspeakers and, thus, increasing the speech intelligibility from multiple loudspeakers. In comparison with ribbed diffusers, the effectiveness of Schroeder diffusers is even greater.

(h)   Strategically located diffusers could be effective for decreasing reverberation, which is useful for increasing speech intelligibility.

(i)   Strategically designed reflectors and obstructions can increase the speech intelligibility from multiple loudspeakers.

Simulation of train noise shows that:

(a)   The overall level of train noise in the area near the end walls is slightly less than that in the other areas.

(b)   Some conventional architectural acoustic treatments in a station are effective when a train is still in the tunnel but not as helpful when the train is already in the station.

(c)   Train noise typically decreases the STI of PA systems from 0·5 to below 0·2 on entering the station.

5.4.   References

5.1   ORLOWSKI R. J. Underground station scale modelling for speech intelligibility prediction. *Proceedings of the Institute of Acoustics (UK)*, 1994, **16**, No. 4, 167–172.

5.2   ORLOWSKI R. J. *London Underground Ltd Research — Acoustic control systems for deep tube tunnels*. Arup Acoustics Report, No. AAc/46318/01/03, 1994.

5.3   KANG J. *Acoustics of long enclosures*. PhD Dissertation, University of Cambridge, England, 1996.

5.4   KANG J. Scale modelling for improving the speech intelligibility of multiple loudspeakers in long enclosures by architectural acoustic treatments. *Acustica/Acta Acustica*, 1998, **84**, 689–700.

5.5   HO C. L. *Station public address — acoustic measurement study*. Hong Kong Mass Transit Railway Corporation Report, ODE/D-TEL/C4246/01, 1992.

5.6   FROMMER G. Technical specification for the consultancy project 'Station acoustics study' (Consultancy No. 92110800-94E). Hong Kong Mass Transit Railway Corporation, 1995/96.

5.7   MAA D. Y. and SHEN H. *Handbook of Acoustics*. Science Press, Beijing, 1987 (in Chinese).

5.8   KANG J. and FUCHS H. V. Predicting the absorption of open weave textiles and micro-perforated membranes backed by an airspace. *Journal of Sound and Vibration*, 1999, **220**, 905–920.

5.9   KANG J. and FUCHS H. V. Effect of water-films on the absorption of membrane absorbers. *Applied Acoustics*, 1999, **56**, 127–135.

5.10   KANG J. Broadening the frequency range of panel absorbers by adding an inner microperforated layer. *Acoustics Letters*, 1998, **21**, 20–24.

5.11   KANG J. *Mehrschichtige Folien Absorber*. Internal report of the Fraunhofer-institut für Bauphysik, Stuttgart, 1993.

5.12   POLACK J. D., MARSHALL A. H. and DODD G. Digital evaluation of the acoustics of small models: The MIDAS package. *Journal of the Acoustical Society of America*, 1989, **85**, 185–193.

5.13   MEYNIAL X., POLACK J. D., DODD G. and MARSHALL A. H. All scale room acoustics measurement with MIDAS. *Proceedings of the Institute of Acoustics (UK)*, 1992, **14**, No. 2, 171–177.

5.14   KUTTRUFF H. Schallausbreitung in Langräumen. *Acustica*, 1989, **69**, 53–62.

5.15   BARNETT P. W. Acoustics of underground platforms. *Proceedings of the Institute of Acoustics (UK)*, 1994, **16**, No. 2, 433–443.

5.16   KANG J. Improvement of the STI of multiple loudspeakers in long enclosures by architectural treatments. *Applied Acoustics*, 1997, **51**, 169–180.

5.17   KANG J. Experimental approach to the effect of diffusers on the sound attenuation in long enclosures. *Building Acoustics*, 1995, **2**, 391–402.

5.18   KANG J. Speech intelligibility improvement for multiple loudspeakers by increasing loudspeaker directionality architecturally. *Building Services Engineering Research and Technology*, 1996, **17**, 203–208.

5.19   KANG J. Effect of ribbed diffusers on the sound attenuation in long enclosures. *Proceedings of the Institute of Acoustics (UK)*, 1995, **17**, No. 4, 499–506.

5.20   SCHROEDER M. R. Diffuse sound reflection by maximum length sequences. *Journal of the Acoustical Society of America*, 1975, **57**, 149–151.

5.21   D'ANTONIO P. and KONNERT J. The reflection phase grating diffuser: design theory and application. *Journal of the Audio Engineering Society*, 1984, **32**, 228–238.

5.22   MILLAND E. *Improvement of the intelligibility of an underground station's public address system*. DEA (Diplôme d'Etudes Approfondis) Thesis, Université du Maine, France, 1994.

5.23   FUJIWARA K. and MIYAJIMA T. Absorption characteristics of a practically constructed Schroeder diffuser of quadratic residue type. *Applied Acoustics*, 1992, **35**, 149–152.

5.24   KUTTRUFF H. Sound absorption by pseudostochastic diffusers (Schroeder diffusers). *Applied Acoustics*, 1994, **42**, 215–231.

5.25   WU T., COX T. J. and LAM Y. W. From a profiled diffuser to an optimized absorber. *Journal of the Acoustical Society of America*, 2000, **108**, 643–650.

5.26   AMERICAN NATIONAL STANDARDS INSTITUTE (ANSI). *Method for the calculation of the absorption of sound by the atmosphere*. ANSI S1.26. ANSI, 1995 (Revised 1999).

5.27   KANG J. Fifty-two years of the auditorium acoustical scale modelling. *Chinese Applied Acoustics*, 1988, **7**, 29–35 (in Chinese).

5.28   KANG J. Modelling of train noise in underground stations. *Journal of Sound and Vibration*, 1996, **195**, 241–255.

5.29   KANG J. Scale modelling of train noise propagation in an underground station. *Journal of Sound and Vibration*, 1997, **202**, 298–302.

6. Speech intelligibility in long enclosures

Due to the special acoustic properties of long enclosures, as indicated in the previous chapters, it is useful to examine various aspects of speech intelligibility in long enclosures. Based on a series of articulation tests in a corridor and a seminar room, this chapter analyses the differences in speech intelligibility between long and regularly shaped (i.e. quasi-cubic) enclosures, between a single loudspeaker and multiple loudspeakers, and the interrelationship between speech intelligibility and the STI in long enclosures. A particular aim of this chapter is to examine the differences and implications of intelligibility between different languages, especially using PA systems in long enclosures. This is because most objective indices for speech intelligibility, such as the STI and the AI, are essentially based on studies of Western languages and not tonal languages like Chinese. Consequently, if the speech intelligibility of an enclosure is satisfactory for English, it is not necessarily satisfactory for Chinese, or vice versa.

6.1. Articulation test

The articulation test was carried out in a long corridor and a regularly shaped seminar room, using loudspeaker sources [6.1,6.2]. Three languages, English and two commonly spoken Chinese dialects, Mandarin and Cantonese, were considered. Briefly, the articulation test was in four steps:

(*a*) selecting comparable test materials for the three languages;
(*b*) recording the test materials on a signal tape in an anechoic chamber;
(*c*) designing experimental conditions; and
(*d*) conducting listening tests.

6.1.1. Test materials

An essential requirement of the articulation test was that the test materials of different languages should be comparable. In English there are four main

Table 6.1. Comparison of articulation test materials between English and Mandarin [6.3–6.5]. In the Mandarin lists, the numbers indicate which tone to use out of four possible tones. In the English PB syllable list the symbols ' and - mean short and long pronunciation, respectively

Mandarin PB words	English PB words	English PB spondees	Mandarin PB syllables	English PB syllables
3bang 4yang	are	airplane	1a	du-
1bo li 1bei	bad	armchair	4an	da'
4ce 2liang	bar	backbone	4ang	bo
4da	bask	bagpipe	3bi	bi
4dian	box	baseball	4bian	gaa
1feng zheng	cane	birthday	1biao	goi
3gan 4mao	cleanse	blackboard	1bin	ju'
2guo 4ji 3zhu 4yi	clove	bloodhound	1ba	pu
2huo 4dong	crash	bonbon	3chi	pow
1jin	creed	buckwheat	1chong	te-

categories of useful test materials for articulation tests, namely nonsense PB syllables, PB monosyllabic words, PB spondaic words and sentences [6.3,6.4]. Similarly, in Mandarin there are PB syllables, PB words and sentences for articulation tests [6.5]. In Cantonese no appropriate test material has been found.

It appears that the English and Mandarin PB words or syllables are not directly comparable. Table 6.1 shows a comparison of some typical PB words and syllables between English and Mandarin [6.3–6.5]. In the Mandarin lists, the numbers indicate which tone to use out of four possible tones. Firstly, the English PB words, especially the monosyllabic ones, only represent the English words with relatively few phonemes and letters, whereas the Mandarin PB words, which include words with one to four Chinese characters, represent all kinds of words in Chinese. In conventional Mandarin PB word lists, the percentages of words with one, two, three and four Chinese characters are 25·5%, 68·7%, 5% and 0·8%, respectively. Secondly, the Mandarin PB syllables, each corresponding to one Chinese character, are mostly meaningful and tonal and, thus, might be understood more easily than the nonsense English PB syllables. Thirdly, if phonetically marked by Chinese characters, English monosyllabic words correspond to one to three Chinese characters. As a result, given the fundamental differences between English and Mandarin, it appears unreasonable to compare the PB word or syllable scores between the two languages.

Alternatively, the comparison of intelligibility between English and Mandarin can be made indirectly. There are two ways of making the indirect comparison. One way is to compare the variation of PB word or syllable scores from one condition to another between the two languages. The other way is to convert the articulation scores obtained with lists of PB words or syllables into sentence intelligibility scores for comparison. The relationships between sentence intelligibility and PB word or syllable scores have been established for both English and Mandarin [6.3,6.5–6.7].

A more direct way to compare the intelligibility of different languages is to use sentences as test materials. The disadvantages of using sentences, however, should also be considered. It has been found that articulation scores obtained with lists of sentences are often so high that the acoustic conditions must differ considerably before a substantial difference in the scores is obtained [6.3]. Moreover, psychological and cultural factors could affect the results.

Accordingly, the articulation tests were carried out using the following test materials:

(a)   PB words in English and Mandarin. For both the languages 25 PB word lists were used. The lists of PB syllables were not chosen due to the high requirement of training for the English testing crew.

(b)   Hong Kong MTR PA messages in English, Mandarin and Cantonese.

There were three reasons for using the PA messages rather than conventional sentence lists. Firstly, in Cantonese no standard sentence list for articulation tests has been found. Secondly, with the PA messages the test materials in the three languages could have the same meaning. Thirdly, since the original MTR PA messages were both in English and Chinese, possible effects caused by translation could be avoided. According to the original PA messages, 25 sentence lists were carefully compiled in the three languages. Each list contained four sentences. To ensure that they would be easy to remember when read to the listeners, the sentence length was never more than 15 words. Some special words that were only familiar to Cantonese listeners, such as the names of MTR stations, were avoided.

6.1.2.   Signal tape

The above test materials were recorded on a signal tape in an anechoic chamber for subsequent presentation. The recordings were made using a tape recorder (NAGRA IV-SJ) with a single omnidirectional microphone. Four native announcers for each language, two males and two females, were selected [6.4]. The announcers were graduate students who have neither obvious speech defects nor noticeable regional accents. They were trained until they were thoroughly familiar with all the test materials and the recording method.

Each list of PB words or PA sentences was divided equally into four sections and read by the four announcers consecutively. To avoid systematic differences between the announcers, the division of PB word lists was in a random rather than alphabetic order. Each PB word was included, without emphasis, in the same carrier sentence, 'would you write ___ now' in English, and '3qing 3xie 4xia ___ 1yi 2ci' in Mandarin. Similarly, before each PA sentence there was a carrier sentence, 'please write the following sentence', to remind the listeners to pay attention to the following sentence. After each PB word or PA sentence there was enough time interval to allow listeners to write the test materials down.

6.1.3. Experimental conditions

Experiments were carried out mainly in a corridor, which is a typical long enclosure. For comparison, a regularly-shaped seminar room was also considered. The design of the experimental conditions was based on two basic requirements:

(*a*) coverage of a considerable range of the STI; and
(*b*) consideration of both reverberation and ambient noise.

Figures 6.1 and 6.2 illustrate the experimental arrangement in the corridor and the seminar room, respectively.

6.1.3.1. Long enclosure

The length, width and height of the corridor (King's College, Cambridge) were 25·2 m, 1·86 m and 2·34 m, respectively. The boundaries of the corridor were acoustically hard and smooth. To avoid possible effects caused by multiple reflections between two end walls, 5 cm-thick foam absorbers were arranged on each end wall at a density of 50%.

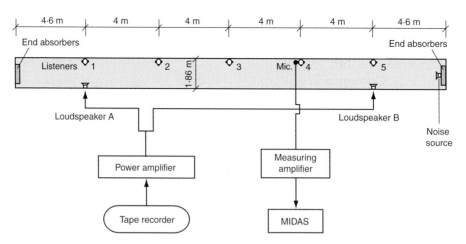

Figure 6.1. Experimental arrangement in the corridor (cases L1, L2 and L3) — plan view

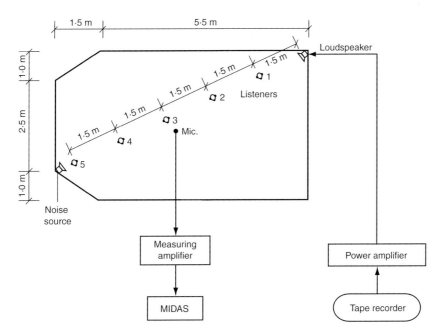

Figure 6.2. Experimental arrangement in the seminar room (cases N1 and N2) — plan view

The sources were loudspeakers (KEF Type 103) on a side wall of the corridor. This was, in principle, to simulate a PA system in some long enclosures. The loudspeaker height was 1·2 m. Since the sound field with a single source could be very different from that with multiple sources (see Section 3.4), experiments were carried out using a single loudspeaker as well as two loudspeakers. In the latter case, in order to obtain a wide range of the STI, the two loudspeakers were arranged symmetrically with a spacing of 16 m. In other words, each loudspeaker was 4·6 m from an end wall. For convenience, the two loudspeakers are called loudspeakers A and B (see Figure 6.1). Correspondingly, five listening positions, namely points 1 to 5 in Figure 6.1, were arranged between the two loudspeakers at intervals of 4 m.

The objective indices in the corridor, such as the STI, reverberation and SPL distribution, were studied carefully using the MIDAS system (see Section 5.1.1). The receiver was a single omnidirectional microphone with a height of 1·2 m. As expected, with loudspeaker A only, the reverberation time increased with increasing source-receiver distance and, correspondingly, the STI decreased continuously from points 1 to 5. Without the effect of increased ambient noise, namely $S/N > 25$ dB(A), from points 1 to 5 the STI became 0·75 from 0·56.

To extend the STI range and investigate the effect of ambient noise, a tape recorder was positioned at one end of the corridor to play white noise (see

Figure 6.1). In principle, this could be regarded as a simulation of train noise in underground stations when the train is in the tunnel. When the S/N ratio caused by this noise source was 9 dB(A) at point 1, the STI variation from points 1 to 5 became 0·60 to 0·19. The STI was calculated by the method given by Houtgast and Steeneken (see Section 1.5). In the calculation the S/N ratios were determined using the average of conventional English and Mandarin spectrums [6.5,6.8].

With the same ambient noise level as above, the STI with two loudspeakers varied between 0·36 and 0·51 over the five listening points. As expected, at point 1 the STI with two loudspeakers was lower than that of loudspeaker A only. The main reason for this was that the EDT of two loudspeakers was longer than that of loudspeaker A alone (see Section 3.4). Another reason was that the power amplifier output was constant for one or two loudspeakers and, thus, at point 1 the S/N ratio of two loudspeakers was slightly lower than that of loudspeaker A only.

The articulation test in the corridor was carried out in the following three cases:

(a)   case L1 — loudspeaker A only, noise source off, S/N > 25 dB(A) at all listening points;

(b)   case L2 — loudspeaker A only, noise source on, S/N = 9 dB(A) at point 1; and

(c)   case L3 — two loudspeakers, same noise level as case L2.

The variations of the STI and EDT along the length in the three cases are shown in Figures 6.3 and 6.4, respectively.

6.1.3.2.   Regularly shaped enclosure

The seminar room (the Martin Centre, Cambridge) was 4·5 m by 7 m by 3·5 m. The EDT of this room was around 0·6–0·8 s at 1000 Hz, as shown in Figure 6.4. A loudspeaker, as used in the corridor, was positioned in a corner of the room at the same height as that in the corridor. Again, the tape recorder was used to play white noise. In correspondence with the corridor, five listening points were arranged in the seminar room. To obtain a wide range of the STI, the listening points were along the room diagonally. Accordingly, the loudspeaker and the noise source were at two diagonal corners of the room.

The articulation test in the seminar room was carried out in the following two cases:

(a)   case N1 — noise source off, S/N > 25 dB(A) at all listening points; and

(b)   case N2 — noise source on, S/N = 2 dB(A) at point 1.

The STIs in the two cases are shown in Figure 6.3.

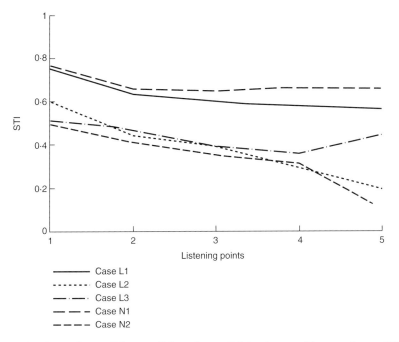

Figure 6.3. STI of case L1, case L2 and case L3 in the corridor, and case N1 and case N2 in the seminar room

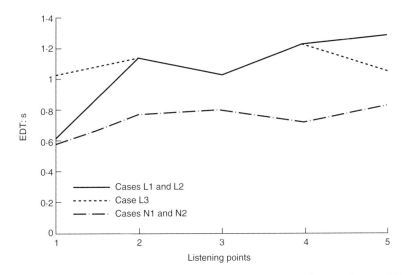

Figure 6.4. EDT of cases L1, L2 and case L3 in the corridor, and cases N1 and N2 in the seminar room

In short, in the two enclosures five acoustic conditions were designed for the articulation test. The five conditions corresponded to 25 STI values with a considerable range of 0·1–0·76.

6.1.4. Listening test

Five native listeners for each language, three males and two females or two males and three females, were selected for the tests. The listeners, who were different from the announcers, were audiologically normal graduate students. For each acoustic condition and each language, five articulation tests were carried out. For English and Mandarin, each test included one PB word list and one PA list. For Cantonese each test included only one PA list. The listeners were required to write down the test material as read to them. Neither PB words nor PA sentences were presented to listeners until the test. For each of the five tests, listeners sat at a different position. In other words, for each of the 25 STIs, articulation scores can be averaged over five listeners. In the corridor, listeners were facing the side wall with loudspeakers. In the seminar room, listeners were facing the loudspeaker. In both enclosures, the ear height was around 1·2 m, which was the same as the loudspeaker height. Full training was given until listeners were familiar with the test method.

In addition to the above articulation tests, a subjective rating was carried out. After each articulation test, listeners were required to give a five-scale rating of the speech intelligibility. The five scales were 1 bad, 2 poor, 3 fair, 4 good and 5 excellent.

For each acoustic condition, the power amplifier output was adjusted so that the signal level was the same for different languages. According to the conventional method, the signal level was measured by $L_{eq}(A)$ of the PA sentences in three lists, namely 12 PA sentences [6.9–6.11].

In brief, from the above tests the following results were obtained for each of the 25 STIs:

(a) word intelligibility in English and Mandarin — percentage of PB words correctly recorded, average of five listeners;

(b) converted sentence intelligibility in English and Mandarin — sentence intelligibility was determined from the above word intelligibility according to the conventional relationship between them [6.3,6.5–6.7, also see Figure 6.6];

(c) PA sentence intelligibility in English, Mandarin and Cantonese — percentage of correctly recorded key words in PA sentences, average of five listeners; and

(d) subjective rating in English, Mandarin and Cantonese — scale point of subjective rating, average of five listeners.

A detailed statistical analysis demonstrated that the discrepancies among listeners and among announcers were acceptable [6.12]. In other words, the number of listeners and announcers used in the test was reasonable.

6.2. Analysis

The analysis that follows is based on the test results and their regressions. Regressions were performed for three different types of relationship between the STI and articulation scores or subjective rating: logarithmic, linear and with a form of second order equation. The logarithmic regressions generally gave the best correlation coefficients r, so these are the ones presented. This reflects the conventional relationship between the STI and articulation scores, which is logarithmic [6.13]. In addition, tests of significance are carried out where appropriate.

6.2.1. Comparison of word and sentence intelligibility among languages

The comparison of word intelligibility between English and Mandarin, as analysed previously, is only relative. Figure 6.5 shows the comparison in the corridor. It can be seen that with a relatively high STI, which corresponds to a large S/N ratio, the word intelligibility of Mandarin is generally better than that of English. This means that the decrease in word intelligibility caused by reverberation is greater in English than in Mandarin. A possible

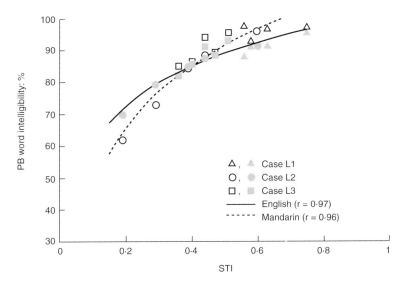

Figure 6.5. Comparison of word intelligibility between English (grey symbols) and Mandarin (open symbols) in the corridor. The curves are logarithmic regressions of the test results

reason for this is that under a reverberant condition some English words, such as affricates, fricatives, nasals and plosives, are very easy to confuse, whereas in Mandarin, tones are helpful for word intelligibility. When the STI becomes lower due to the decrease of S/N ratio, the contrary occurs, the word intelligibility of English becomes better than that of Mandarin. Typical examples are at points 4 and 5 in case L2 (see Figures 6.1, 6.3 and 6.5). It has been observed that in English the SPL dynamic range is considerably greater than that in Mandarin, i.e., typically, for a given L_{eq} (A) the peak level of English is about 3 dB(A) higher than that of Mandarin. This may be an explanation of the fact that, in noisy conditions, some English words may be understood by picking up only the high peaks.

To compare the converted sentence intelligibility between English and Mandarin, it is useful to analyse the relationship between word and sentence intelligibility. Such relationships in English and Mandarin have been found from the literature, as shown in Figure 6.6 [6.3,6.5–6.7]. It is interesting to note that for a given word intelligibility, the sentence intelligibility of English is better generally than that of Mandarin. An important reason for this difference, as indicated previously, is that the Mandarin PB words represent all kinds of Chinese words, whereas the English PB words only represent relatively short English words. Given the fact that the word intelligibility improves with increasing number of sounds per word [6.3], the difference between the two curves in Figure 6.6 is reasonable.

Correspondingly, by converting the word intelligibility in Figure 6.5 into sentence intelligibility according to Figure 6.6, the differences between English and Mandarin become more obvious, as shown in Figure 6.7. In the figure,

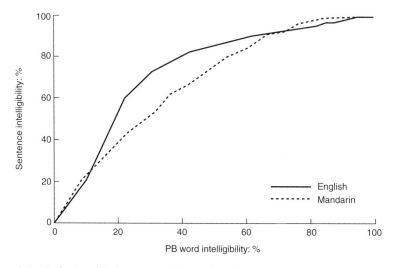

Figure 6.6. Relationship between PB word and sentence intelligibility in English and Mandarin [6.3,6.5–6.7]

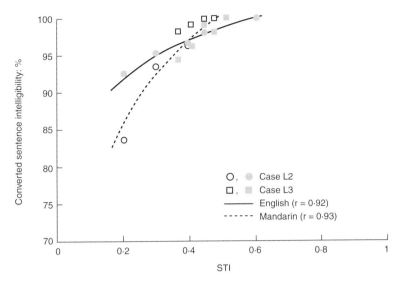

Figure 6.7. Comparison of converted sentence intelligibility between English (grey symbols) and Mandarin (open symbols) in the corridor. The curves are logarithmic regressions of the test results

case L1 is not included since, for both English and Mandarin, the converted sentence intelligibility is above 99% at all listening points.

Similarly, in the corridor the differences in PA sentence intelligibility between English and Mandarin, as shown in Figure 6.8, are also systematic, especially in cases L2 and L3, in which the noise source was used. By comparing a typical PA sentence in English and Mandarin, 'please use the queuing lines and let passengers leave the train first' and '3qin 2pai 4dui, 1xian 4rang 2cheng 4ke 3xia 1che', it is clear that the total number of sounds in Mandarin is significantly less than that in English. This might be another reason for the lower sentence intelligibility of Mandarin in comparison with English. The disadvantage of fewer sounds in Chinese, however, appears to be compensated by other factors. It is interesting to note in Figure 6.8 that with a relatively low STI, which corresponds to a low S/N ratio, the PA sentence intelligibility of Cantonese is considerably higher than that of Mandarin and close to that of English. This is probably because in comparison with Mandarin, in Cantonese there are more tones (nine) and the SPL dynamic range is greater.

Generally, in the seminar room the differences among the three languages are similar to those in the corridor. Figures 6.9 and 6.10 show the comparison of word intelligibility and converted sentence intelligibility between English and Mandarin, respectively. Again, case N1 is not included in Figure 6.10 due to the high sentence intelligibility in both languages. It can be seen that in case N2, in which the STI was relatively low due to the low S/N ratio, the

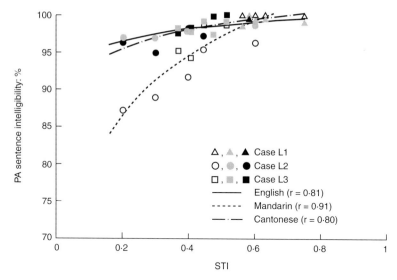

Figure 6.8. Comparison of PA sentence intelligibility among English (grey symbols), Mandarin (open symbols) and Cantonese (solid symbols) in the corridor. The curves are logarithmic regressions of the test results

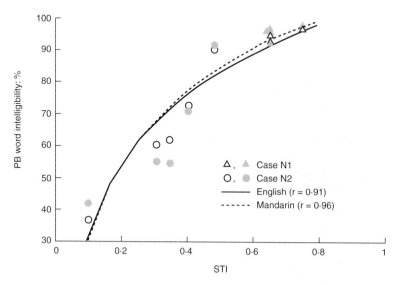

Figure 6.9. Comparison of word intelligibility between English (grey symbols) and Mandarin (open symbols) in the seminar room. The curves are logarithmic regressions of the test results

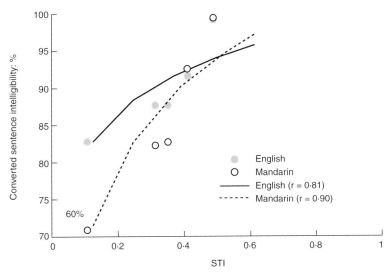

Figure 6.10. Comparison of converted sentence intelligibility between English and Mandarin in case N2 of the seminar room. The lines are logarithmic regressions of the test results

sentence intelligibility of English is obviously better than that of Mandarin (see Figure 6.10), although the difference in word intelligibility between the two languages is not as systematic as in the corridor (see Figure 6.9). Similarly, in case N2 the differences in PA sentence intelligibility between English and Mandarin are also considerable, as shown in Figure 6.11. Corresponding to Figure 6.8, the PA sentence intelligibility of Cantonese is better than that of Mandarin and close to that of English. Unlike the results in the corridor, in case N1 of the seminar room, in which the STI was relatively high due to the high S/N ratio, the differences in word and PA sentence intelligibility among the three languages are unnoticeable (see Figures 6.9 and 6.11). This is probably due to the short reverberation time in this room.

In summary, the above comparisons suggest that in terms of speech intelligibility, Mandarin is slightly better than English under reverberant conditions, English is considerably better than Mandarin under noisy conditions and Cantonese is better than Mandarin and close to English under noisy conditions.

The differences in intelligibility among other languages have also been noticed. Houtgast and Steeneken, when demonstrating the effectiveness of the RASTI with ten Western languages, indicated that language-specific effects could be a factor causing disparity among various tests [6.9]. Differences between Japanese and Western languages have also been reported [6.14,6.15].

In a number of languages, relationships between objective indices, such as the STI, and various measures of intelligibility, such as sentence intelligibility, have been established. In order to compare the difference in intelligibility

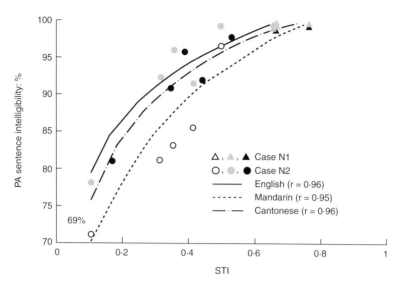

Figure 6.11. Comparison of PA sentence intelligibility among English (grey symbols), Mandarin (open symbols) and Cantonese (solid symbols) in the seminar room. The curves are logarithmic regressions of the test results

between the languages for a given objective index, however, it appears to be inappropriate to directly compare these relationships because they were established under different conditions. Moreover, these relationships are mainly for sound fields that are close to diffuse and may not be applicable for non-diffuse fields [6.14,6.16], such as in long enclosures.

6.2.2. Comparison of subjective rating among languages

It appears that the above differences among the three languages are unnoticeable in terms of subjective rating. Figures 6.12 and 6.13 show the subjective ratings of the three languages in the corridor and the seminar room, respectively. It can be seen that in both enclosures there is no systematic difference in subjective rating among the three languages. Under the noisy condition, despite the higher sentence intelligibility in English, the subjective rating given by English listeners is not higher, and sometimes even lower, than that by Mandarin listeners. Similar differences between subjective ratings and articulation scores can also be seen by comparing the results of Mandarin and Cantonese.

The above results suggest that the subjective rating type of test could be rather misleading. However, since a low subjective rating may correspond to less confidence in speech intelligibility, the factor of subjective appraisal should also be considered in the acoustic design for an enclosure used for multiple languages.

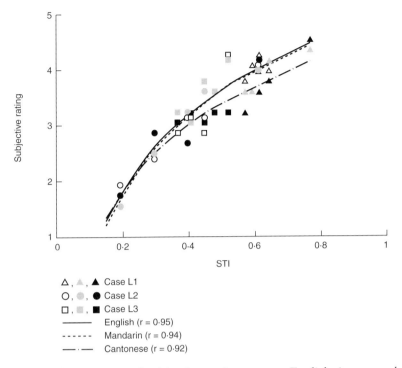

Figure 6.12. Comparison of subjective rating among English (grey symbols), Mandarin (open symbols) and Cantonese (solid symbols) in the corridor. The curves are logarithmic regressions of the test results

6.2.3. Comparison between one and two loudspeakers

It is interesting to note that for a given STI in the corridor, the intelligibility of speech caused by a single loudspeaker and two loudspeakers is not the same. By comparing case L3 with the other two cases in Figures 6.5, 6.7 and 6.8, it can be seen that for all the three languages, with two loudspeakers the word or sentence intelligibility is slightly higher than that with a single loudspeaker. The difference is about 3–5% in word intelligibility and 1–4% in sentence intelligibility. A t-test shows that these differences are significant (word intelligibility, $p < 0.001$; sentence intelligibility, $p < 0.01$). This appears to indicate that the intelligibility of sound arriving from two widely separated directions is better than that of sound from one direction.

6.2.4. Comparison between two enclosures

Since the sound fields in long and regularly shaped enclosures are rather different, as analysed in Chapter 2, it is useful to compare the speech intelligibility between the two enclosures above. Figure 6.14 compares the word

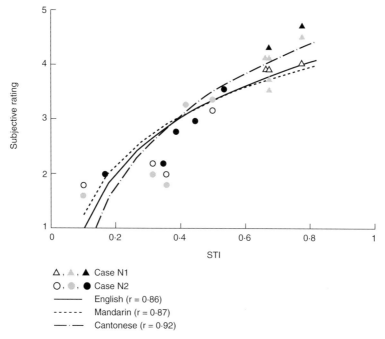

Figure 6.13. Comparison of subjective rating among English (grey symbols),
Mandarin (open symbols) and Cantonese (solid symbols) in the seminar room.
The curves are logarithmic regressions of the results

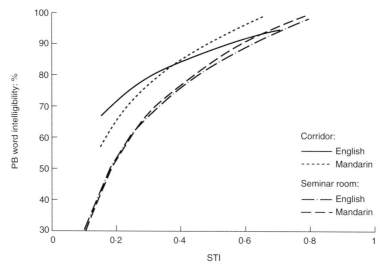

Figure 6.14. Comparison of word intelligibility between the three cases in the
corridor and the two cases in the seminar room. The regression curves are the
same as those in Figures 6.5 and 6.9

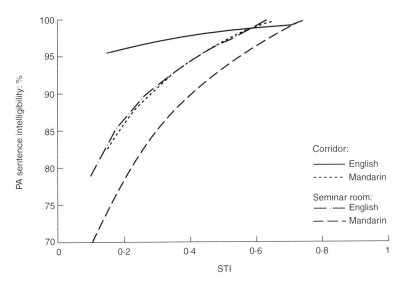

Figure 6.15. Comparison of PA sentence intelligibility between the three cases in the corridor and the two cases in the seminar room. The regression curves are the same as those in Figures 6.8 and 6.11

intelligibility, for both English and Mandarin, between the regression of the 15 tests (cases L1, L2 and L3) in the corridor and the regression of the ten tests (cases N1 and N2) in the seminar room.

Correspondingly, the comparison of PA sentence intelligibility between the two enclosures is shown in Figure 6.15. It can be seen that for a relatively high STI there is no systematic difference between the two enclosures, whereas for a relatively low STI, the word or sentence intelligibility in the seminar room is noticeably lower than that in the corridor. This difference is likely to be due to the difference in sound field between the two enclosures. For example, in the corridor the noise was mainly from one direction (see Figure 6.1) and, thus, could be less disturbing than that in the seminar room.

In contrast to the word or sentence intelligibility, the difference in subjective rating between the two enclosures appears to be insignificant. This can be seen by comparing Figures 6.12 and 6.13.

6.2.5. Correlation between speech intelligibility and the STI in the corridor

Generally speaking, in the corridor the STI is highly correlated to the articulation scores as well as the subjective rating, as shown in Figures 6.5, 6.7 and 6.8, although the relationships are not necessarily the same as those in regularly shaped enclosures like the seminar room. The correlation coefficient between the STI and word intelligibility is 0·97 for English and 0·96 for Mandarin.

This means the STI, which has been proved to be effective for regularly shaped enclosures, is also useful for long enclosures.

6.3. Summary

The articulation test has shown some special features in long enclosures in terms of speech intelligibility:

(*a*) With a given STI, the word and sentence intelligibility in the long enclosure was noticeably higher than that in the regularly shaped room when the S/N ratio was relatively low.

(*b*) For a given STI in the long enclosure, with two loudspeakers the word and sentence intelligibility was slightly higher than that with a single loudspeaker.

(*c*) In the long enclosure the STI was highly correlated to the intelligibility of speech.

Differences in intelligibility between languages have been demonstrated. The results suggest that in terms of speech intelligibility, Mandarin is slightly better than English under reverberant conditions, English is considerably better than Mandarin under noisy conditions, and Cantonese is better than Mandarin and close to English under noisy conditions. This means that for a space that is used for different languages, the difference in intelligibility between them should be considered in the acoustic design. In underground stations, where speech intelligibility is significantly affected by the background noise, a higher STI may be required if Mandarin is included. Conversely, the differences in subjective rating among the three languages appear to be unnoticeable. This should also be taken into account in the acoustic design.

6.4.   References

6.1   KANG J. Comparison of speech intelligibility between English and Chinese. *Journal of the Acoustical Society of America*, 1998, **103**, 1213–1216.

6.2   KANG J. Intelligibility of public address system: A comparison between languages. *Proceedings of the 7th West Pacific Conference on Acoustics (WESTPRAVII)*, Kumamoto, Japan, 2000, 145–148.

6.3   BERANEK L. L. *Acoustic Measurements*. John Wiley and Sons, New York, 1949.

6.4   AMERICAN NATIONAL STANDARDS INSTITUTE (ANSI). *Method for measuring the intelligibility of speech over communication systems*. ANSI S3.2. ANSI, 1989.

6.5   MAA D. Y. and SHEN H. *Handbook of Acoustics*. Science Press, Beijing, 1987 (in Chinese).

6.6   EGAN M. D. *Architectural Acoustics*. McGraw-Hill, New York, 1988.

6.7   TEMPLETON D. (ed.) *Acoustics in the Built Environment*. Butterworth Architecture, Oxford, 1993.

6.8   INSTITUTE OF BUILDING PHYSICS. *Handbook of Architectural Acoustics*. China Building
        Industry Press, Beijing, 1985 (in Chinese).
6.9   HOUTGAST T. and STEENEKEN H. J. M. A multi-language evaluation of the RASTI-
        method for estimating speech intelligibility in auditoria. *Acustica*, 1984, **54**, 185–199.
6.10   STEENEKEN H. J. M. and HOUTGAST T. RASTI: a tool for evaluating auditoria. *Brüel
        & Kjær Technical Review*, 1985, **3**, 13–39.
6.11   STEENEKEN H. J. M. and HOUTGAST T. Subjective and objective speech intelligibility
        measures. *Proceedings of the Institute of Acoustics (UK)*, 1994, **16**, No. 4, 95–111.
6.12   KANG J. *Acoustics of long enclosures*. PhD Dissertation, University of Cambridge,
        England, 1996.
6.13   HOUTGAST T. and STEENEKEN H. J. M. The modulation transfer function in room
        acoustics. *Brüel & Kjær Technical Review*, 1985, **3**, 3–12.
6.14   NOMURA H., MIYATA H. and HOUTGAST T. Speech intelligibility and modulation
        transfer function in non-exponential decay fields. *Acustica*, 1989, **69**, 151–155.
6.15   UCHIDA Y., LILLY D. J. and MEIKLE, M. B. Cross-language speech intelligibility
        in noise: the comparison on the aspect of language dominance. *Journal of the
        Acoustical Society of America*, 1999, **106**, 2151.
6.16   YEGNANARAYANA B. and RAMAKRISHNA B. S. Intelligibility of speech under non-
        exponential decay conditions. *Journal of the Acoustical Society of America*, 1975,
        58, 853–857.

Index

Page numbers in italics refer to figures and tables.